Smart Sensors, Measurement and Instrumentation

Volume 44

Series Editor

Subhas Chandra Mukhopadhyay, School of Engineering, Macquarie University, Sydney, NSW, Australia

The Smart Sensors, Measurement and Instrumentation series (SSMI) publishes new developments and advancements in the fields of Sensors, Instrumentation and Measurement technologies. The series focuses on all aspects of design, development, implementation, operation and applications of intelligent and smart sensors, sensor network, instrumentation and measurement methodologies. The intent is to cover all the technical contents, applications, and multidisciplinary aspects of the field, embedded in the areas of Electrical and Electronic Engineering, Robotics, Control, Mechatronics, Mechanical Engineering, Computer Science, and Life Sciences, as well as the methodologies behind them. Within the scope of the series are monographs, lecture notes, selected contributions from specialized conferences and workshops, special contribution from international experts, as well as selected PhD theses.

Indexed by SCOPUS and Google Scholar.

Shantanu Pal · Zahra Jadidi · Ernest Foo ·
Subhas C. Mukhopadhyay
Editors

Emerging Smart Technologies for Critical Infrastructure

Springer

Editors
Shantanu Pal
School of Information Technology
Deakin University
Melbourne, VIC, Australia

Zahra Jadidi
School of Information and Communication Technology
Griffith University
Brisbane, QLD, Australia

Ernest Foo
School of Information and Communication Technology
Griffith University
Brisbane, QLD, Australia

Subhas C. Mukhopadhyay
School of Engineering
Macquarie University
Sydney, NSW, Australia

ISSN 2194-8402 ISSN 2194-8410 (electronic)
Smart Sensors, Measurement and Instrumentation
ISBN 978-3-031-29844-8 ISBN 978-3-031-29845-5 (eBook)
https://doi.org/10.1007/978-3-031-29845-5

This Springer imprint is published by the registered company Springer Nature Switzerland AG
The registered company address is: Gewerbestrasse 11, 6330 Cham, Switzerland

Preface

With the proliferation of smart sensors, wearable sensor-rich devices, advanced communication and networking technologies, Internet of things (IoT), artificial intelligence, machine learning, blockchain, and digital twins, enable the continuing development of existing technology to a newer extent for critical infrastructure. In this context, the protection of smart critical infrastructure is highly demanding. This includes preventing unauthorized users and devices from accessing the network and application security that must be placed on hardware and software to lock down possible vulnerabilities. Emerging vulnerabilities in smart critical infrastructure in industry 4.0 can change the nature of the connected devices on how they operate IoT applications and how they communicate with one another. The emerging trends in IoT and related materializing technologies are driven by artificial intelligence, blockchain, cloud, fog, edge, and communication technologies for short range to long range as well as for resource-constraint devices (e.g. memory capacity, processing power, battery, etc.) must be considered.

Innovations in emerging technologies to promote their benefits in numerous applications and services in smart critical infrastructure including health care, power grid, environmental pollution monitoring, smart transportation, industrial automation, education, agricultural sectors, etc. However, numerous challenges need to be addressed to provide better and more efficient applications in future. Some of these challenges are: (i) lack of IoT technologies and standards for business processes, (ii) limited guidance for life cycle maintenance and management of resource constrained IoT devices, (iii) available best practices for the AI and blockchain developers, (iv) security, privacy, and trust issues in general, (v) data analytics tools for analysis of huge data volumes, etc. New challenges and opportunities demand a fertile ground for research and innovation to develop and deploy emerging technologies in the broad context of societal applications. This requires novel methodological, theoretical, and mathematical modelling, implementing protocols, and computational methods that reflect recent advancements.

The complete book is composed of seven chapters. These chapters contain a wide range of information from novel infrastructures, architectures, and protocols to theories, services, and architectures for smart critical infrastructure using emerging technologies.

The first chapter "Cybersecurity for Satellite Smart Critical Infrastructure" discusses the satellite infrastructure and the cybersecurity frameworks applied in smart critical infrastructure. It identifies three main cybersecurity properties for satellite smart critical infrastructure: real-time analysis, mitigation mechanism, and low computational overhead. These properties are mapped against existing cybersecurity frameworks applied in smart critical infrastructure. The examination results indicate that the existing cybersecurity frameworks need to be more inapplicable, incompatible, or inadequate to address the cyber-attacks in satellite smart critical infrastructure. In addition, the chapter highlights a combination of mechanisms, e.g. runtime verification and digital twin technology, to address the satellite smart critical infrastructure cybersecurity.

The second chapter "Blockchain in Smart Grids: A Review of Recent Developments" provides a systematic and update-to-date review of blockchain integration in smart critical infrastructure, e.g. smart grids. First, the chapter comprehensively introduces blockchain and smart grids and outlines a taxonomy for the application sub-areas of blockchain in smart grids. Then, it reviews the state of the art in each sub-area based on the available literature to provide an insightful understanding of how blockchain can support the operation of modern smart grids. Finally, the chapter delivers a valuable and motivational reference to using blockchain for smart critical infrastructure.

The third chapter "Client Selection Frameworks Within Federated Machine Learning: The Current Paradigm" presents a systematic discussion of the client selection frameworks within Federated Machine Learning (FML). This chapter discusses the significance of investigating ways to utilize big data further and the benefits of this for smart critical infrastructure. Furthermore, it identifies the challenges, e.g. computation cost and privacy issues in traditional machine learning, and focuses on addressing them using FML. Finally, the chapter emphasizes a better way to operate a client selection framework for smart critical infrastructure.

The fourth chapter "Explainable Anomaly Detection in IoT Networks" provides the significance of security monitoring in Cyber-Physical System (CPS) networks and the issues related to the increasing number of threats against these CPS networks. The chapter indices the limitations of machine learning methods used to analyse network data and detect intrusions automatically from a black box point of view that has no explanation for their decision. The chapter highlights the need for explainable machine learning techniques to explain the reasons behind the decision made by machine learning-based intrusion detection systems (IDSs). Finally, a NetFlow-based analysis is discussed, which is a scalable method suitable for a high volume of data.

The fifth chapter "Application of Machine Learning on Material Science and Problem Solving Under Security—A Review" systematically discusses machine learning applications in material science and problem-solving under security. The chapter is presented into two parts applicable to machine learning models, (i) material science and (ii) problem-solving. The chapter aims to introduce automated solutions for material science, known as material informatics, to prevent information security threats, particularly information integrity threats that are probable in national security organizations. Then the chapter attends to two applicable graph-based models in problem-solving, namely graph-cut models (deterministic) and a unified graphical model (probabilistic) in a simple word to address unsolving problems, e.g. computer vision, engineering, security, and medicine that are critical in smart critical infrastructure.

The sixth chapter "Introduction to Blockchain Technology with Bitcoin Protocol" emphasizes the need for engineering technology, e.g. the blockchain that powers bitcoin, and its attention as a potential software solution for various industrial applications. The authors examine the fundamental design principles of blockchain technology using bitcoin's architecture as a foundation and understand the rationale behind its design and the limitation of scalability.

The seventh chapter "Security Challenges and Wireless Technology Choices in IoT-Based Smart Grids" discusses essential background information on the residential smart grid system, one of the smart critical infrastructures. The chapter identifies security risks and attacks that threaten the smart grid systems. A systematic review and discussion of relevant modern IoT transmission technologies covering their benefits, key performance metrics, and their appropriate place within the IoT-enabled smart grid systems are presented. Finally, a discussion on security challenges and future research directions addressing the identified security challenges in IoT-enabled smart grid systems are presented in detail.

We sincerely appreciate Deakin University, Griffith University, and Macquarie University. We would also like to acknowledge the contributions of all the researchers who contributed to the chapters of this book.

Melbourne, Australia

Dr. Shantanu Pal
Dr. Zahra Jadidi
Prof. Ernest Foo
Prof. Subhas C. Mukhopadhyay

Contents

Cybersecurity for Satellite Smart Critical Infrastructure

Ayodeji James Akande, Ernest Foo, Zhe Hou, and Qinyi Li

Abstract A satellite communication system, as a typical example of the Internet of things, is a smart critical infrastructure and has become an essential component used in various services such as finances, communications, ground and air-borne navigation, utilities, power grid distribution, emergency services, agriculture, banking, and many other critical industries. In recent times, satellite communication systems have become a target for cyber-attack. In this chapter, we review satellite infrastructure and the existing cybersecurity frameworks applied in smart critical infrastructure. We identified three main cybersecurity properties for satellite smart critical infrastructure, which are real-time analysis, mitigation mechanism, and low computational overhead. These properties are mapped against existing cybersecurity frameworks applied in smart critical infrastructure. The result indicated that the existing cybersecurity frameworks are either inapplicable, incompatible, or inadequate to address the cyber-attacks in satellite smart critical infrastructure. In addition, we identify a combination of mechanisms such as runtime verification and digital twin technology to address the satellite smart critical infrastructure cybersecurity. Finally, we discuss a review of the mechanisms and their applications along with our future work.

Keywords Runtime verification · Satellite smart critical infrastructure · Digital twins · Static verification · Secure time synchronization protocol · Cyber security

A. J. Akande (✉) · E. Foo · Z. Hou · Q. Li
Griffith University, Nathan, QLD 4111, Australia
e-mail: ayodeji.akande@griffithuni.edu.au

E. Foo
e-mail: e.foo@griffith.edu.au

Z. Hou
e-mail: z.hou@griffith.edu.au

Q. Li
e-mail: qinyi.Li@griffith.edu.au

S. Pal et al. (eds.), *Emerging Smart Technologies for Critical Infrastructure*,
Smart Sensors, Measurement and Instrumentation 44,
https://doi.org/10.1007/978-3-031-29845-5_1

1 Introduction

A satellite system is a smart critical infrastructure that is used in various essential services such as finances, communications, ground and air-borne navigation, utilities, power grid distribution, emergency services, agriculture, banking, and many other critical industries. In recent years, there has been an increase in cyber attacks on satellite communications. According to Graczyk et al. [27], the dependence on space made it an important asset and a worthwhile target for protection.

The satellite communication system is designed to be smart with features and modules for efficient communications. However, the major challenges facing the protection of space satellite systems include the isolated nature of the deployed satellite and the high latency and high error environment that communications must travel in. The other challenge is the limited processing capacity on board the satellite, as power is mainly provided by a solar battery. These challenges thereby create an unprecedented complexity. According to Falco [21], due to the cost and accessibility to some features, the ground station functionality historically has been exclusively afforded by select nation-states, but the recent introduction of cloud-based ground stations for satellite control has provided unprecedented access to the services. "Coupled with low-cost CubeSats that are rife with cybersecurity issues, it is now feasible for a wide range of nation-states, companies, or even individuals to cause harm to other satellites in orbit" [21]. The nature of the satellite system results in difficulties in anticipating all possible faults/hazards, vulnerabilities, and cyber-attacks.

Some of the common cyber attacks on satellite communications include spoofing and jamming of satellite communication attacks. Spoofing attacks on satellite communications can provide access to data that can be used by adversaries to cause serious evoke. Likewise, the jamming of satellite communications can create a great negative impact. To address the challenges facing satellite communication systems, a technique is required which will involve real-time monitoring without affecting the system's operations or affecting computational power and timely response to anticipated faults/hazards, vulnerabilities, malicious events or cyber-attack.

Asides from confidentiality, integrity, availability, authentication and authorisation, there is a need to set properties for a cybersecurity framework that analyses the satellite smart critical infrastructure's applications or programs and operations. The cybersecurity framework must ensure; first, to identify malicious events and/or anticipate faults or possible cyber-attacks; secondly, to mitigate against those events in a timely manner; and thirdly, must not add computational overhead to the system. Several cybersecurity frameworks have been developed by researchers and applied in various industries and fields. These frameworks are either incompatible, inapplicable or inadequate to provide the desired cybersecurity for a satellite smart critical infrastructure. However, there are existing technologies and security mechanisms that can be combined to design a suitable cybersecurity framework for satellite communication systems, such as digital twin technology and runtime verification.

Around the world, a lot of research has been performed on digital twin technology. This is a technology developed to imitate a real system. The digital twin is a virtual representation of a physical object or process that serves as the real-time digital counterpart. Digital twin-based satellite communication can be developed to run computational heavy tasks to monitor and verify satellite communication system performance and efficacy. With the concept of the digital twin, it is possible to model, anticipate all possible faults/hazards and be proactive to stop unwanted events before they happen if events happening at runtime can be analysed. Though the digital twin supposes to replicate the real system, a verification technique is essential to analyse the system's program as being executed, monitor the results of the execution and analyse the outcome to find anomalies and security breaches. The exploration of formal verification, specifically runtime verification, provides the technique. Runtime verification "is one of the efficient ways to monitor and verify the security of satellites and other space assets" [32]. According to Goldberg et al. [26], "NASA has developed a runtime verification technique that can be applied to check autonomous agents running on the PLASMA planning system".

Digital twin verification using runtime verification relies on the correctness of data and fresh state data. Though "the correctness of data can be achieved via standard data integrity/authentication techniques such as message authentication codes and digital signatures, the latter is particularly challenging considering that the satellite communication suffers from unpredictable delays" [32]. To address this issue, Hou et al. [32] focused on establishing a proper level of time synchronisation among digital twins. However, the time synchronisation only addressed the delay in satellite communications, while verification of the consistency of satellite application/program is not yet addressed. In addressing this issue, runtime verification with properties expressed in linear temporal logic can be implemented. System applications must set and meet properties such as configuration values.

Though the concept of runtime verification of digital twins-based satellite communication systems has not been fully researched, this paper will review research work on the concept and its applications in various fields.

Our Literature Review Approach In this paper, research was conducted using scholarly search engines "Scopus" and "Google Scholar". The review covers various areas such as satellite system cybersecurity, cybersecurity frameworks in smart critical infrastructure, and the runtime verification-digital twins concept. The literature review approach is made up of three stages: (a) relevant keywords search from databases, (b) exclusion of irrelevant papers by reading abstracts, (c) full-text reading of relevant papers, and classification of papers according to the properties presented in Sect. 3. Databases, search strings, and time scope of the literature review were used.

Our literature review is in two phases; the first phase is to review satellite infrastructure, identify the existing cybersecurity frameworks and evaluate their suitability for satellite infrastructure cybersecurity while the second phase is to review mechanisms suitable to form a framework for satellite infrastructure cybersecurity. Shown in Fig. 1 is a graph indicating a growth rate in the satellite critical infrastructure cybersecurity research field from 2008 till the present.

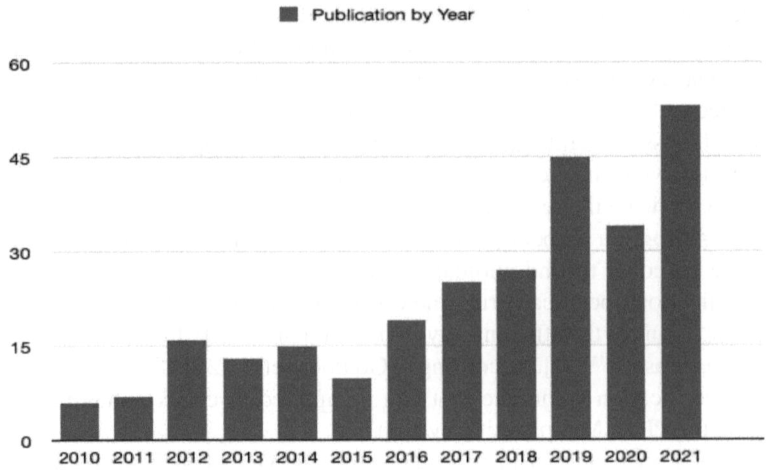

Fig. 1 The number of publications per year based on the research topic "Cybersecurity framework for smart critical infrastructure"

In Sect. 2, we present the literature review of satellite smart critical infrastructure and list out its infrastructural limitation for cybersecurity analysis. Section 3 discusses cybersecurity properties for satellite critical infrastructure and also reviews the existing framework against the identified properties, while presented in Sect. 4 are the mechanisms that provide satellite smart critical infrastructure cybersecurity and their existing areas of applications. Section 5 is the conclusion of this paper, and future research work is highlighted.

2 Survey of Cybersecurity Properties for Satellite Smart Critical Infrastructure

The satellite infrastructure as a space asset has become a critical component used in various essential services such as finances, communications, ground and air-borne navigation, utilities, power grid distribution, emergency services, agriculture, banking, and many other critical industries. Since the 80s, Space assets has moved from being only used by the military and are now increasingly used by civilian. Some basic satellite system components include a communication system, sensors, actuators, onboard computer system, and power [38].

As shown in Fig. 2, satellite infrastructure has two main components: the ground station, which consists of fixed or mobile transmission, reception, and ancillary equipment, and the space satellite. Both space satellites and ground stations transmit via uplink and downlink channels. A ground station transmits the signal from the earth's space to a satellite. The satellite receives, amplifies the signal, and re-transmits it back to Earth, where it is received. The received signal is then re-amplified by ground stations.

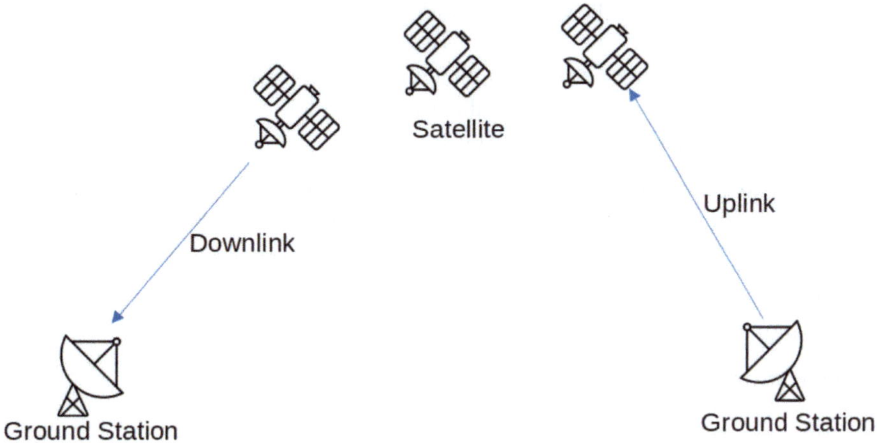

Fig. 2 A simple communication between the satellite station and the ground station

Kim [38] stated that "satellites are usually equipped with a kind of payload system(s) (radio/TV transmitter/transducer, radar, telescope or different scientific instrument, etc.) to perform certain dedicated space mission(s)" [38]. Kim [38] further identified some types of satellites which include navigation, communication, Earth observation, scientific, geophysics, geodetics, technology demonstration, and developers training.

"Ubiquitous use of Global Navigation Satellite Systems (GNSS), including Global Positioning System (GPS) in civilian, security, and defence applications, and the growing dependence on them within critical infrastructures has highlighted the need for protection against vulnerability due to intentional or unintentional interference sources" [6]. According to Graczyk et al. [27], the dependence on space made it an important resource and a worthwhile target for protection. "Unfortunately, several threats put the sustainable use of space at risk, both as a foundation for military operations, but more importantly for the economic applications that affect our daily life" [27]. Therefore, a cybersecurity framework must be designed against cyber-attacks in satellite critical infrastructure.

Nweke [45] identified two components, namely the CIA model and AAA. The CIA model describes important goals of cybersecurity while AAA describes a method through which cybersecurity is achieved. Satam et al. [51] stated that it is unrealistic to apply encryption techniques to sensors due to their low (or no) computational power but proposed the authentication of sensors and their data. However, authentication may not be sufficient for a satellite smart critical infrastructure cybersecurity.

Several cybersecurity frameworks have been developed by researchers and applied in various industries and fields. These frameworks include the deployment of intrusion detection systems [18, 46], blockchain technology [1], Software Defining Network (SDN) technology [31] and STRIDE threat modeling [24]. The existing cybersecurity frameworks are either inapplicable, incompatible, or inadequate for satellite smart critical infrastructures cybersecurity due to the following:

1. the isolated nature of the deployed satellite,
2. the high latency,
3. high error environment that communications must travel in,
4. the limited processing capacity onboard of the satellite, and
5. the low computational power of sensors.

In order to design a suitable framework for satellite infrastructure in the space environment, some properties need to be considered. These properties are:

- Low computational overhead
- Real-time detection
- Mitigation mechanism.

In the next section, we will discuss the properties in detail.

3 Cybersecurity Properties for Satellite Smart Critical Infrastructure and Definitions

A cybersecurity framework is a mechanism designed to ensure adequate protection of a system against cyber-attacks. Due to the limitations as listed in Sect. 2, three cybersecurity properties have been identified for satellite infrastructure. These properties will be used to develop a suitable framework.

Property 1 (Low Computational Overhead) *Every analytical process must require minimum computational overhead without affecting the operation of the system except when the analytical process can be performed externally.*

Due to the low computational power of onboard sensors, any monitoring and analysis technique for satellite cybersecurity is either of low computational overhead or can be implemented outside the space environment.

In addressing the low computational overhead problem, an engineering technology such as digital twin technology can be implemented. As discussed in Sect. 4.1, with Digital Twin (DT) technology, a replica of the satellite infrastructure can be developed, which serves as the real-time digital counterpart of the physical system or process.

Satellite data analysis will be a computationally heavy task and can be performed on high-performance computers on the ground. Hence, the idea of using a digital twin, which is a virtual representation that serves as the real-time digital counterpart of a physical object or process.

The digital twin reflects the real-time status of the physical twin, it is natural to ask whether we can analyze events happening at runtime and be proactive to stop bad things before they happen.

Property 2 (Real-Time Detection) *Detection of undesirable events must be timely and preferably performed at run-time.*

Due to the limited processing capacity onboard the satellite and to proffer a solution to the high error environment that satellite communication travel in, real-time analysis of satellite data is quite challenging to be performed on onboard satellite infrastructure.

Real-time detection is the use of data and related resources for the analysis and discovery of malicious events as soon as it enters the system. This involves real-time monitoring of every process in the satellite communication system, which includes monitoring data exchange in space missions. The real-time analysis process entails the extraction of data from a running system and the use of the data to identify behaviours satisfying or violating the defined security measures. The real-time analysis enables data scientists or analysts to use analytical data for forming operational decisions and applying them, displaying ongoing operations with constantly updated transactional data sets and reporting historical and current data simultaneously. A real-time monitoring tool consists of an aggregator that gathers data events from a variety of data sources and an analytic engine that analyzes the data, correlates values, and blends streams together.

Real-time analysis onboard satellite infrastructure seems challenging. To provide that support for real-time detection, the use of digital twins with formal verification techniques such as runtime verification can be implemented. Detection of malicious events is not sufficient to protect critical infrastructure, mitigating against cyber-attack is highly important to avert the impact on services.

Property 3 (Mitigation) *Every process must be verified, and if the process does not satisfy the defined security property for the system, the process will be prevented from further action.*

While real-time analysis detects anomalies or events suggesting cyber-attacks, mitigation techniques react to the observed behaviours violating set properties. Due to the isolated nature of the deployed satellite, a mechanism to promptly respond to the cyber-attack and anticipated fault. An effective and efficient cybersecurity mechanism for satellite infrastructure must proactively mitigate unwanted events before happening.

Due to the limitation in processing capacity, real-time mitigation onboard satellite infrastructure seems challenging. To provide mitigation, runtime verification with temporal logic can be implemented. Runtime verification, as discussed in Sect. 4.2 is an efficient formal verification technique to monitor and verify the security of satellites and other space assets. Using temporal logic, security properties can be defined in the runtime verification algorithm to validate every process of the system.

Also, due to addressing the high latency, a secure time synchronization protocol is required. Time synchronization between the space satellites and the ground station relies on the communication network. Satellite communication experiences high latency and is likewise prone to cyber attacks due to the isolated nature of the deployed satellite and the high error environment that communications must travel in. As the correctness of data is highly important, it is essential for a verifier to know the freshness of the state data. The attacks may sabotage the time synchronization

Table 1 The number of papers on the existing cybersecurity frameworks addressing the our properties

Properties	No of papers
Real-time analysis	105
Mitigation process	23
Computational power	0

by preventing the packets from being correctly transmitted, rendering inaccurate synchronization, and further making the runtime verifier receives and checks an outdated or manipulated state. Thus, time synchronization protocols must be sufficiently robust against active adversaries. A secure state synchronization method to ensure secure and efficient long-distance communication between the satellite and ground station is important for cybersecurity. The verifier must know the freshness of the data state. To provide secure time synchronization between the satellite and ground station, Authenticated Network Time Protocol (ANTP) can be deployed. It is a better alternative to the traditional time synchronization protocol, Network Time Protocol because it offers security.

As shown in Table 1, out of 128 documents, 105 documents discussed the real-time analysis in their framework without clearly indicating mitigation process [11, 35, 43] while 23 stated mitigation approaches against the detected attacks [2, 16, 31, 57, 61]. However, none of the framework implementations described in these papers will be able to address the computational overhead problem in satellite infrastructure. The implementation of these security frameworks may increase computational power in satellite critical infrastructure making them not fit to be used for satellite communications cybersecurity.

We further reviewed the top 13 publications and shown in Table 2, is the analysis of the frameworks mapped against our cybersecurity properties for satellite smart critical infrastructure. Though the combination of digital twins, authenticated network time protocol (ANTP), and runtime verification is still a very young idea that has only been briefly explored in the analysis of cyber-physical systems recently, this chapter reviews the application of digital twins, ANTP and runtime verification in various smart critical infrastructure systems including satellites and space missions. There is a significant lack of technical details in the literature.

4 Mechanisms for Satellite Smart Critical Infrastructure Cybersecurity

Protecting smart critical infrastructure such as satellite systems from all anticipated/possible faults/hazards, vulnerabilities, and cyber-attacks has become a great concern due to the system's complexity and heterogeneity, and integration with the

Table 2 The mapping of the existing framework against our three properties for the satellite cybersecurity framework

References	Application	Real-time	Mitigation	Comp. overhead
[7]	Satellite system	N	N	Y
[3]	Satellite system	Y	Y	Y
[1]	Transportation	Y	Y	Y
[31]	Transportation	Y	Y	Y
[57]	Transportation	Y	Y	Y
[61]	Transportation	Y	Y	Y
[24]	Transportation	Y	N	Y
[49]	Power and energy	Y	Y	Y
[51]	Smart infrastructures	Y	N	Y
[35]	Smart infrastructures	Y	N	Y
[33]	Power and energy	Y	N	Y
[18]	Smart city	Y	N	Y
[46]	Smart city	Y	N	Y

While "Y" is yes and "N" is no, "Real-time" represents the framework that addresses real-time data analysis and detection, "Mitigation" represents the framework that addresses the response mechanism to detected malicious events while "Comp. overhead" represents the framework that if implemented in satellite infrastructure will increase the onboard computational power

Internet. A satellite system can be classified as an example of cyber-physical infrastructure. Research has indicated that "with the rise of new technology trends, such as AI Foundations, Intelligent Things, Cloud to Edge, or Immersive Experiences, many of today's paradigms can be expected to be disrupted" [30]. Box [14] stated that the Cyber-Physical Systems (CPS) complexity made it "practically impossible to give accurate models, enumerate all use-cases, and to anticipate all possible faults/hazards during development" [14].

Falco [21] presented satellite-to-satellite attacks. The paper described a class of satellite-to-satellite cyber attacks and explained that the attacks were previously limited to a select group of nation-states, but the low-cost CubeSats and ground stations, along with cloud services make the system accessible to adversaries and cyber attacks are increasingly feasible [21]. The paper explained that an attack could be performed without typical housing on satellites. An offensive satellite with special-purpose sensors and actuators can be used to perform a cyber attack. These actuators can be controlled via ground station or decision-system algorithms resident on the satellite's onboard computer systems. The satellite communication system can be attacked by an adversary via some of its components such as sensors, actuators, onboard computer systems, communication systems, and protocols. However, an adversary will need to learn about the whereabouts of the satellite, and Falco [21] presented two ways that an adversary can determine the location of its victim; by "using local proximity sensors or by collecting information from a third-party system" [21]. The paper described the attacks against the satellite components as "complex and may require near-field or line-of-sight proximity to the targeted asset" [21]. It further stated that to ensure the protection of the satellite system against manipulation, a robust ground station control with near real-time capabilities for signal delivery and processing will be required.

In another paper by Amin et al. [6], two main threats were identified and analyzed namely, jamming interference and spoofing attacks. Spoofing attacks on satellite communications can provide access to data collected and logged by the satellite, which can be used by adversaries to cause serious evoke. Likewise, the jamming of satellite communications can create a great negative impact. For example, financial institutions depend on satellite communication to provide precise timing for high-speed trading while coordinating signal handshakes and enabling connectivity, wireless networks, and cellphone towers rely on satellite communication timing. Therefore, a breach of such services may lead to catastrophic events.

Some of the possible satellite failures include actuator failures such as the Canadian telecommunication satellites Anik E1 and Anik E2 in 1994, which were "caused by an electrostatic discharge in both satellites disrupting the momentum wheel control" [23]. Another example of satellite failure is onboard computer system (OBCS) failure. A typical example was the detection of mission-threatening anomalies with the Attitude Determination and Control System (ADCS) from the Challenger spacecraft by NASA's Tracking and Relay Data Satellite (TRDS) 1 launched in 1983 [62]. "This was caused by Single Event Upsets (SEUs) "orbit flipping" that yielded state changes in random access memory on the onboard computer system" [62]. Other satellite failures include power system failure, sensor failure [12], and communication system failure [19].

Modern satellite communication system designs are smart with features and modules for efficient communications and also for a secure onboard analysis, resulting in unprecedented complexity and heterogeneity, making the protection of deep space satellites challenging.

In developing a framework suitable for satellite smart critical infrastructure cybersecurity, one of the useful mechanisms is engineering technology such as digital twin technology.

4.1 Digital Twin Technology

Digital twin (DT) is the virtual representation of a real system, digital replicas of actual physical systems (living or not), interweaving solutions of complex systems analysis, decision support, and technology integration. A digital twin replicates an object or system that spans its life-cycle, updated from real-time data, and to help decision-making, utilizing simulation, machine learning, and reasoning [37, 44, 60]. The literature review indicated that digital twin has been applied in many fields along with satellite communication, though little work has been done in using digital twin to model satellite communication for cyber attack analysis.

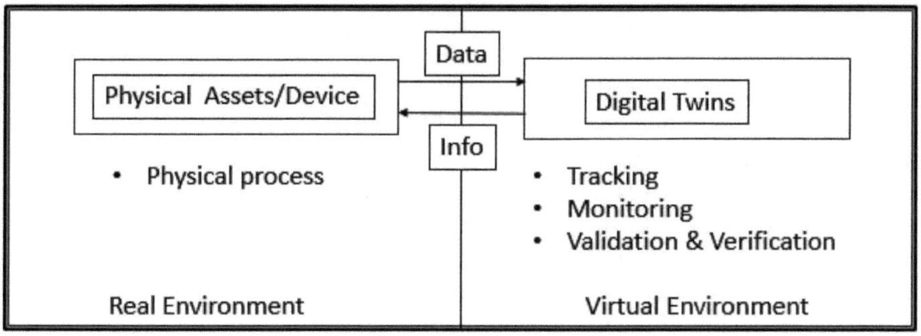

Fig. 3 A simple representation of digital twin technology

As shown in Fig. 3, the digital twin is a virtual representation of a physical environment and forms its processes from data obtained from the real system. The digital twin components information includes necessary action to be taken by the real system. In the case of digital twin-based satellite communication, computational heavy tasks can be performed on a computer system at the ground station which serves as the digital twin.

In 2003, Grieves proposed the concept for industrial product lifecycle management [28]. Grieves [28] described the DT technique from three aspects: a physical entity, a virtual entity, and a data connection. The concept of DT has gained much attention in both academia and the industry due to its benefit and application potential. "Digital twins (DT) are increasingly adopted by several disciplines, including the manufacturing [39], automotive [15] and energy sectors [54], agriculture [48], aerospace engineering, robotics, smart manufacturing, renewable energy, and process industry [28, 29]".

Glaessgen and Stargel [25] "described the digital twin as an integrated multiphysics, multi-scale, probabilistic simulation of a complex product and uses the best available physical models, sensor updates, etc., to mirror the life of its corresponding twin". DT have been useful for converging the physical and virtual spaces [58], guaranteeing information continuity through the system lifecycle [53], system development and validation through simulation [13], and preventing undesirable system states [29].

Flammini [22] worked on data-driven evaluation and prediction of critical dependability attributes such as safety and introduced a conceptual framework based on autonomic systems to host DT run-time models based on a structured and systematic approach. The paper argued that "the convergence between DT and self-adaptation is the key to building smarter, resilient and trustworthy CPS that can self-monitor, self-diagnose and ultimately self-heal". Flammini [22] associated the concepts of resilience, self-healing, and trustworthy autonomy with the paradigm of DT through run-time models embedded in the MAPE-K loop of autonomic computing. Discussed in the paper was "an overview of the main concepts and their interrelations as well as some reference abstract models and architectures for continuous CPS monitoring for faults and anomalies using DT and self-healing mechanisms" [22]. In another paper by Jiang et al. [34], the industrial applications of Digital Twins were presented. The paper discussed the challenges in industrial practice today and described step by step how the identified challenges could be addressed using Digital Twins techniques. Jiang et al. [34] reviewed the current industrial practice, which indicated that several desirable objectives could not easily be achieved. The paper identified five unachievable desirable objectives amidst others, and they include "the optimization of the design outcome because the field practices of manufacturing are not taken into account during design creating a gap between design and manufacturing, the prediction of the quality of the product during manufacturing, introducing another gap between manufacturing and inspection, the improvement of the design and manufacturing processes by learning from the previous batches of the manufactured goods, a gap in product batches, and the inability to control the fluctuations in plant-wide cost-there is a lack of a real-time information thread across different stages of the product lifecycle" [34].

In research by Errandonea et al. [20], a literature review on the use of digital twins for maintenance was presented. The paper focused on the review of DT applications for maintenance in various industrial sectors in several application areas such as design, production, manufacturing, and maintenance. The researcher stated that one of the benefits offered by DT is intelligent maintenance strategies. The paper explained the concept of "Digital Twin" and "maintenance".

Some researchers presented literature views on the application of DT in various sectors and fields for maintenance [41, 47]. Pedro [47] stated that DT is "a key enabler for efficient verification and validation processes, stressing out the importance of its own validation and accreditation phase" [47]. Also, Löcklin et al. [41] presented a survey of approaches that use Digital Twins for verification and validation purposes. The paper investigated the application of the Digital Twin for verification and validation, ranging from the validation of non-functional properties to the verification of safety-critical requirements.

Digital Twin for Satellite Systems Shangguan et al. [52] introduced a new physical-virtual convergence approach, digital twin, for fault diagnosis and health monitoring (FD-HM) applicable in satellite systems. Shangguan et al. [52] mentioned that data-driven Fault Diagnosis and Health Monitoring (FD-HM) approaches had been developed using signal processing or data mining to extract implicit information from the operating state of the system useful for monitoring the system. The paper, however, highlighted the limitation of the approaches. "These approaches for the FD-HM of the satellite system are driven primarily by the historical data and some static physical data, with little consideration for the simulation data, real-time data, and data fusion between the two, so it is not fully competent for the real-time monitoring and maintenance of the satellite in orbit" [52]. Shangguan et al. [52] also presented an FD-HM application of the satellite power system to demonstrate the effectiveness of the proposed approach.

In another related work, Yang et al. [63] explored the method of constructing DT of spacecraft. The paper identified that "the spacecraft is facing more frequent and multi-task tests in an unprecedented complex environment and the current challenge lies in how to further build an integrated system of the virtual and physical space for spacecraft" [63]. Yang et al. [63] presented the concept of Spacecraft Digital twin (SDT) and four stages of simulation development from DT's perspective. Moreover, in the research work was the proposal of the conceptual structure of a four-dimensional model to adapt spatial distribution.

In another work by Liu et al. [40], Global Navigation Satellite System (GNSS) was merged with Digital Twins (DTs) techniques to address tedious controlling practices in building operation and maintenance (O&M) processes. The paper emphasized that controlling building operation and maintenance (O&M) processes require extensive visualization and trustworthy decision-making strategies, which could not effectively be achieved with existing technologies and practices. The paper presented a method for achieving intelligent control of building O&M processes relaying on Global Navigation Satellite System (GNSS) with Digital Twins (DTs) techniques. Global Navigation Satellite System (GNSS) was utilized to capture real-time building information during building O&M processes.

With the concept of the digital twin, which reflects the real-time status of the physical twin, it is possible to observe and maintain the system's operation. However, there is a need to ensure the accuracy and correctness of the execution of the program on critical systems and to anticipate all possible faults/hazards if events happen at runtime. Therefore, the use of formal verification can be explored, specifically runtime verification.

4.2 *Formal Verification for Space Assets*

Formal verification is the process of proving or checking the correctness of a program/system, and it could be static or runtime. There are various formal verification methods, such as static verification and runtime verification. Static verification verifies properties of all possible runs of a program while runtime verification monitors the execution of a system, detecting violations as they appear at runtime [5].

Ahrendt et al. [5] highlighted the difference between static and runtime verification. Static verification may be more effective and efficient, but the techniques "either have high precision, in which case powerful judgments are hard to achieve automatically, or they use abstractions supporting increased automation, but possibly losing important aspects of the concrete system in the process" [5]. On the other hand, Runtime verification combines full precision of the model (including the real deployment environment) with full automation, but its limits include the inability to judge future, alternative runs and the computational overhead of monitoring the running system, which may not be typically high but can still be prohibitive in certain settings [5].

In another paper, Ahrendt et al. [4] proposed a framework to combine both static analysis techniques and runtime verification. The proposed framework is based on a suitable combination of static and dynamic verification techniques, in particular, based on the underlying approaches of the deductive theorem prover KeY and the runtime verification tool Larva. The paper explained that even though static verification of software has become more relevant, effective and efficient, there are some inherent limitations. Despite static verification providing high precision, in some "cases powerful judgments are still too hard to achieve automatically, while others use abstractions to enable increased automation, in which case important, or even critical, aspects of the real, concrete system are easily missed, not to speak of the fundamental difficulty of crafting the right abstraction" [5]. To address the limitations, there is a need for lightweight formal methods such as runtime verification, which are easier to exploit but give limited guarantees. The paper presented "the conceptual model of a framework for the verification of object-oriented systems and proposed ppDATEs as a unified specification language for describing both static and dynamic properties, and demonstrated an example to illustrate how the approach could be used" [5]. The authors also described two application domains that could benefit from the approach: Electronic and legal contracts; and Transaction-handling systems.

Runtime Verification Runtime verification is a formal verification approach that analyzes programs as they are executed, monitors the results of the execution, and uses analyzed results to find anomalies and security breaches. Runtime verification increases standards system compliance, that is, it verifies when the execution of the program is not in compliance if properties set for the program are not met. Runtime verification as a program tracks the execution errors that traditional testing or static analysis may not find.

Bartocci et al. [8] presented a brief introduction to the field of runtime verification, and it covered four major areas: "how to specify system behaviour, how to set up monitoring, how to perform instrumentation, and what the limitations of monitoring are" [8]. Runtime verification analyses the execution of the system and not its code and rigorously detects bugs or errors while scaling to large code bases, unlike traditional formal analysis techniques, like model checking or deductive verification. Runtime verification performs synchronous monitoring in which the system does not proceed further until it is confirmed that the action did not violate the specification.

In a related work by Luppen et al. [42], a case study in formal specification and runtime verification of a CubeSat communications system was presented. Specifications to detect and trigger appropriate mitigation for CubeSat communications system faults were designed. The research work identified that the commonplace for CubeSat projects are failed communications to the ground stations, and to address the issue, a mechanism must be in place that will be able to detect faults in a CubeSat's communications system, which will aid in preventing a premature mission end. The Realizable, Responsive, Unobtrusive Unit (R2U2) tool was deployed within CubeSat communications systems leveraging on runtime verification technique. Luppen et al. [42] developed "a reference set of formal specifications in mission-time linear temporal logic (MLTL) describing a modelled CubeSat communications system, detailed the validation strategy over these specifications using experimental evaluation with the R2U2 tool, discussed specification patterns that emerge while developing and revising the runtime verification specifications and presented the lessons learned from validating the specifications that may inform future CubeSat runtime verification efforts" [42].

Runtime Verification with Digital Twins Few pieces of research have been done to demonstrate the use of runtime verification with Digital twins. Kang et al. [36] proposed a novel framework, DigTwinOps (Digital Twin framework for Operation of Cyber-Physical Production Systems). This is a Digital Twin framework for Runtime Verification of Cyber-Physical Production Systems (CPPSs), which provides runtime controllability verification of a control command of a CPPS application. As explained in the paper, "DigTwinOps manages the ECML-based Digital Twin Model that synchronizes the states of real machines in the production environment and provides monitoring and simulation services to both CPPS application and human worker for verifying the controllability of the decided control action" [36]. According to the paper, a high-level structure of the existing production system was modelled as a digital twin, and to perform monitoring and simulation, the framework of DigTwinOps was used. As stated in the paper, "the framework allows interworking simulations of data from existing factory hierarchies and can be reflected in decision making based on the simulation results of possible control commands" [36].

Saratha et al. [50] presented a paper on a Digital Twin with Runtime-Verification for Industrial Development-Operation Integration. The paper gave an overview of a data model of a digital twin for industrial development-operation integration (DevOps). Using the data model, models are built from development and links models with data from the operation. The paper explained that "the models from development are represented by ontologies that describe the functional decomposition in parts and associated properties, and the properties are linked with symbolic reachability information that is created during development which can be used as a basis for runtime verification" [50]. Using a water level monitor as a case, the experiments indicated that the method for runtime verification could find discrete and parametric faults swiftly and without the need for previous fault modelling.

Likewise, Sleuters et al. [55] described a method to develop digital twins for large-scale distributed IoT systems to address the verification and validation challenges of an operational IoT system. Sleuters et al. [55]'s research was built on Verriet et al. [59]'s work which described virtual prototyping of large-scale IoT control systems using domain-specific languages (DSLs). Sleuters et al. [55] discussed how the virtual prototype generated from the models was connected to the physical system and created a digital twin. However, the paper did not describe how the digital twin created can be used in runtime verification.

Runtime verification depends on defining certain properties for monitoring and analysis of program execution. The properties are verified against the execution of a program to track for the execution errors that traditional testing or static analysis may not find. Runtime verification can verify general properties automatically, requiring no development input, and can also check any specific properties formally defined using certain languages such as temporal logic. Runtime verification programs should ensure that the defined property is not violated.

Bauer et al. [9] examined linear temporal logic derived for finite traces for runtime verification and studied variants of Linear Temporal Logic (LTL)s. For runtime verification logics, the paper considered a linear temporal logic interpreted over finite traces with semantics showing that of LTL over infinite traces. Three existing LTLs interpreted over finite traces were recalled, namely, Fluent Linear Temporal Logic (FLTL), $LTL_{\mp}$, and LTL_3. The properties of the LTL variants were also explained in the paper. Also, four maxims considered to be essential for a LTL purposely for runtime verification were examined which were: "first, existential next requires the inclusion of a strong next operator; second, complementation by negation requires that a negated formula evaluates to the complemented and different truth value; third, impartiality requires that a finite trace is not evaluated to ($\perp$) if there still exists an infinite continuation leading to another verdict; and finally, anticipation requires that once every infinite continuation of a finite trace leads to the same verdict, then the finite trace evaluates to this very same verdict" [9]. These maxims were analyzed against FLTL, $LTL_{\mp}$, and LTL_3 and the result indicated that none of them satisfied all of the four maxims. Therefore, the paper proposed runtime verification linear temporal logic (RV-LTL), whose semantics combines ideas present in LTL_3 as well as FLTL. Furthermore, the paper stated that the "semantics of RV-LTL indicates whether a finite word describes a system behaviour which either satisfies the monitored property, violates the property, will presumably violate the property or will presumably conform to the property in the future, once the system has stabilized" [9]. In the paper, some basic properties of RV-LTL were analyzed, and verified that RV-LTL acts on the four maxims. Furthermore, Bauer et al. [9] developed a monitor generation procedure that relies on corresponding monitor constructions for FLTL and LTL_3 to make RV-LTL a practically applicable device for runtime verification.

In another related paper by Bauer et al. [10], a study of runtime verification of properties expressed either in linear time temporal logic (LTL) or timed linear time temporal logic (TLTL) was presented. The approach is said to be suitable for monitoring discrete-time and real-time systems. The work considered a "finite trace as the incrementally observed finite prefix of an unknown infinite trace" in runtime verification, and depending on the verifying property with the observed prefix, the continuation of the trace may cause the evaluation of the correctness property to either true, false or inconclusive [10]. The paper proposed a three-valued semantics (with truth values true, false, inconclusive) for LTL and for the formulae of the logic, a conceptually simple monitor generation procedure was given, optimally in two respects: "First, the size of the generated deterministic monitor is minimal, and, second, the monitor identifies a continuously monitored trace as either satisfying or falsifying a property as early as possible" [10]. The same road map, that is, three-valued semantics, was proposed for real-time Timed Lineartime Temporal Logic (TLTL) but the corresponding construction of a timed monitor is more involved.

4.3 Authenticated Network Time Protocol

Runtime verification of digital twins relies on correct and fresh state data. While the correctness of data can be achieved via standard data integrity/authentication techniques (e.g., message authentication codes and digital signatures), the latter is particularly challenging considering that satellite communication suffers from unpredictable delays. Thus, establishing a proper level of time synchronisation among digital twins is vital.

Traditionally, time synchronization is implemented using the Network Time Protocol (NTP). This is to synchronize device time with remote servers. However, network time protocol (NTP) does not offer a reasonable level of security against active attacks. Dowling et al. [17] introduced a new authenticated time synchronization protocol called ANTP. Authenticated Time Synchronization Protocol (ANTP) is designed to securely synchronize the time of a client and server, using the public key infrastructure. The protocol was designed "to allow a server to perform a single public key operation per client during the infrequently performed key exchange phase and then use only faster symmetric key operations for each subsequent time synchronization request from that client" [17]. According to the paper, ANTP has been designed to the throughput phase by a factor of only 1.6$\times$ when compared to NTP. For load-balancing purposes, ANTP servers sharing the same long-term secret are designed to handle different phases of the same client. Dowling et al. [17] further explained that for large-scale deployments, ANTP is designed to reduce server-side public key operations by intermittently performing a key exchange using public key cryptography, then relying solely on symmetric cryptography for subsequent time synchronization requests; moreover, it does so without requiring server-side per-connection state. Additionally, ANTP ensures that authentication does not degrade the accuracy of time synchronization. We measured the performance of ANTP by implementing it in OpenNTPD using OpenSSL.

Spanghero et al. [56] presented authenticated time for detecting GNSS attacks. The paper discussed an approach that leverages time obtained over networks a mobile device can connect to, to detect discrepancies between the GNSS provided time and the network time. Spanghero et al. [56] proposed a framework that utilized the ubiquitous IEEE 802.11 (Wi-Fi) infrastructure together with the network time servers. The framework "supports application-layer, secure and robust real-time broadcasting by Wi-Fi Access Points (APs), based on hash chains and infrequent digital signatures verification to minimize computational and communication overhead, allowing mobile nodes to efficiently obtain authenticated and rich time information as they roam" [56]. The framework also includes the pairing of the ubiquitous IEEE 802.11 (Wi-Fi) infrastructure and the network time servers with Network Time Security (NTS), to enhance resilience through multiple sources.

4.4 Integrated Systems for Runtime Verification of Space Assets via Digital Twin

Hou et al. [32] implemented the concept of Runtime Verification with Digital Twins for Space Assets. The concept was designed to provide trustworthy and secure communication for satellites. Though runtime verification is not meant to replace traditional formal verification, runtime verification is suitable for cyber attack assessment in a digital twin satellite system. This is because it is computationally cheaper than other formal verification methods such as model checking and theorem proving. Rather than running a model checking program that analyses all executions of a given system to answer whether it satisfies a given correctness property φ, runtime verification only checks for the word problem. Also, runtime verification deals with online monitoring, which enables the management of its processes incrementally. These features serve as advantages in the implementation of runtime verification with a digital twin satellite system. Runtime verification can be used to confirm the behaviour of the satellite by verifying state information. A runtime verification framework was developed that supports multiple temporal logics such as FLTL and PTLTL in one package [32]. The framework is driven by a model checker tool called Process Analysis Toolkit (PAT). The runtime verification engine for the digital twin can verify properties in the temporal logic languages.

A digital twins system for satellite systems was designed with a focus on security monitoring and verification. The satellite digital twins model consists of both the physical twin which is a satellite station and the digital twin located at the ground station. The physical twin periodically synthesises engineering data into scientific data such as processed datasets, which in our framework contain the states to be synchronized and checked, and sent to the ground station. The scientific data are transmitted to the ground station using our delay-tolerant communication protocol. In the model, the digital twin at the ground station simulates the state of the satellite using processed data and performs computationally heavy tasks while in the satellite digital twins model, the physical twin runs on the actual space asset and collects, monitors, and interprets engineering data such as control and sensor raw data. The digital twin models two essential aspects of the satellite: the physical behaviour, captured by sensor data, and the communications, captured by transmitted messages.

The state of both the physical and digital twins is synchronized using a secured time synchronization protocol, Authenticated Network Time Protocol (ANTP). Authenticated Network Time Protocol uses message authentication codes (MACs) as the only cryptographic tool to provide authenticity and is robust against active adversaries. Once an accurate time synchronization is established, the subsequent network packets containing state information can be time-stamped and then authenticated to allow the verifier to decide if the received state information is fresh for runtime verification. The secure time synchronization protocol protects against malicious entities that are not part of the digital twin system. To improve efficiency, Hou et al. [32] simplify the cryptographic algorithm and key establishment of the ANTP protocol by letting the MAC algorithm store the keys for the MAC in the satellite and the ground station.

5 Conclusion

In this chapter, we presented a review of existing cybersecurity frameworks, evaluated them against the satellite infrastructure and developed three properties required for the satellite infrastructure cybersecurity framework. We analysed the existing framework against the properties and the analysis revealed that these frameworks are either incompatible, inadequate or suitable for satellite infrastructure cybersecurity. We presented a literature review on the identified mechanisms.

Given the identified mechanism for satellite smart critical infrastructure cybersecurity framework, which combines digital twin technology, runtime verification and secure time synchronization protocol, we plan to build on the integration of the mechanisms and perform penetration testing and attack simulation for satellites in the future. We choose to build the runtime verifier on PAT instead of using an existing one or developing one from scratch because we plan to formally verify the entire satellite communication system, which includes satellite behaviour, synchronisation protocol, and digital twins system, among others, and PAT will be used to model and verify many components of the system.

As part of our plan, a large amount of data will be generated while performing penetrating testing, including the status of the system and the parameters of the attacker. From the data, we can use the clustering technique to obtain the states of the agents and learn how their states evolve. With the states and their transitions, we will model the behaviour of agents in Markov decision processes and use reinforcement learning to train the agents towards optimal policies: the most efficient hacks for the attacker and the best counteractions for the defender. With the AI-based simulations, we expect to check more corner cases that might have been missed by human attackers.

Moreover, we plan to model the correct behaviour of space assets, then express desired properties of the system as reachability, deadlockfreeness, liveness, or temporal logic formulae and then verify those properties using PAT. Another test of our approach includes running the proposed framework in a simulation environment based on Gilmore Space Technologies' Electrical Ground Support Equipment satellite simulator. The new simulation environment will re-implement both the satellite and the digital twin, as well as their communication methods.

References

1. Abdel-Basset M, Moustafa N, Hawash H, Razzak I, Sallam K, Elkomy O (2022) Federated intrusion detection in blockchain-based smart transportation systems. IEEE Trans Intell Transp Syst 23(3):2523–2537. https://doi.org/10.1109/TITS.2021.3119968, https://www.scopus.com/inward/record.uri?eid=2-s2.0-85118530257&doi=10.1109%2fTITS.2021.3119968&partnerID=40&md5=13afe899e8ec5d5d7fee1b4f88aefd36

2. Abu Al-Haija Q, Al Badawi A, Bojja G (2022) Boost-defence for resilient IoT networks: a head-to-toe approach. Expert Syst. https://doi.org/10.1111/exsy.12934, https://www.scopus.com/inward/record.uri?eid=2-s2.0-85122150135&doi=10.1111%2fexsy.12934&partnerID=40&md5=0161e1e317766828c84976847ddff2f3
3. Adhikari S (2020) Building cyber resilience in space assets with real-time autonomous graph database anomaly detection algorithms. In: ASCEND 2020, p 4113
4. Ahrendt W, Chimento JM, Pace GJ, Schneider G (2015) A specification language for static and runtime verification of data and control properties. In: International symposium on formal methods. Springer, pp 108–125
5. Ahrendt W, Pace GJ, Schneider G (2012) A unified approach for static and runtime verification: framework and applications. In: International symposium on leveraging applications of formal methods, verification and validation. Springer, pp 312–326
6. Amin MG, Closas P, Broumandan A, Volakis JL (2016) Vulnerabilities, threats, and authentication in satellite-based navigation systems [scanning the issue]. Proc IEEE 104(6):1169–1173
7. Ashraf I, Narra M, Umer M, Majeed R, Sadiq S, Javaid F, Rasool N (2022) A deep learning-based smart framework for cyber-physical and satellite system security threats detection. Electronics 11(4):667
8. Bartocci E, Falcone Y, Francalanza A, Reger G (2018) Introduction to runtime verification. In: Lectures on runtime verification. Springer, pp 1–33
9. Bauer A, Leucker M, Schallhart C (2010) Comparing LTL semantics for runtime verification. J Logic Comput 20(3):651–674
10. Bauer A, Leucker M, Schallhart C (2011) Runtime verification for LTL and TLTL. ACM Trans Softw Eng Methodol (TOSEM) 20(4):1–64
11. Bhardwaj A, Kumar M, Stephan T, Shankar A, Ghalib M, Abujar S (2022) IAF: IoT attack framework and unique taxonomy. J Circuits Syst Comput 31(2). https://doi.org/10.1142/S0218126622500293, https://www.scopus.com/inward/record.uri?eid=2-s2.0-85117486347&doi=10.1142%2fS0218126622500293&partnerID=40&md5=092187a421ed89496b19b893c08afdb0
12. Board DMI (2006) Overview of the DART mishap investigation results. Technical report. NASA. Available at http://www.nasa.gov/pdf
13. Boschert S, Rosen R (2016) Digital twin? The simulation aspect. In: Mechatronic futures. Springer, pp 59–74
14. Box GE (1976) Science and statistics. J Am Stat Assoc 71(356):791–799
15. Caputo F, Greco A, Fera M, Macchiaroli R (2019) Digital twins to enhance the integration of ergonomics in the workplace design. Int J Ind Ergon 71:20–31
16. Chattopadhyay A, Lam KY, Tavva Y (2021) Autonomous vehicle: security by design. IEEE Trans Intell Transp Syst 22(11):7015–7029. https://doi.org/10.1109/TITS.2020.3000797, https://www.scopus.com/inward/record.uri?eid=2-s2.0-85118825860&doi=10.1109%2fTITS.2020.3000797&partnerID=40&md5=5aca4524e7c0e21145ba6766beafa74d
17. Dowling B, Stebila D, Zaverucha G (2016) Authenticated network time synchronization. In: 25th USENIX security symposium (USENIX security 16), pp 823–840
18. Elsaeidy A, Elgendi I, Munasinghe KS, Sharma D, Jamalipour A (2017) A smart city cyber security platform for narrowband networks. In: 2017 27th international telecommunication networks and applications conference (ITNAC). IEEE, pp 1–6
19. Enayat M (2012) Satellite jamming in Iran: a war over airwaves. Small Media Report, Kasım
20. Errandonea I, Beltrán S, Arrizabalaga S (2020) Digital twin for maintenance: a literature review. Comput Ind 123:103316
21. Falco G (2020) When satellites attack: satellite-to-satellite cyber attack, defense and resilience. In: ASCEND 2020, p 4014
22. Flammini F (2021) Digital twins as run-time predictive models for the resilience of cyber-physical systems: a conceptual framework. Philos Trans R Soc A 379(2207):20200369
23. Galvan DA, Hemenway B, Welser I, Baiocchi D et al (2014) Satellite anomalies: benefits of a centralized anomaly database and methods for securely sharing information among satellite operators. Technical report. RAND National Defense Research Institute, Santa Monica, CA

24. Girdhar M, You Y, Song T, Ghosh S, Hong J (2022) Post-accident cyberattack event analysis for connected and automated vehicles. IEEE Access 1. https://doi.org/10.1109/ACCESS.2022.3196346, https://www.scopus.com/inward/record.uri?eid=2-s2.0-85135752860&doi=10.1109%2fACCESS.2022.3196346&partnerID=40&md5=0e55cc365258f5e0205fd198d8a61ddf
25. Glaessgen E, Stargel D (2012) The digital twin paradigm for future NASA and US air force vehicles. In: 53rd AIAA/ASME/ASCE/AHS/ASC structures, structural dynamics and materials conference 20th AIAA/ASME/AHS adaptive structures conference 14th AIAA, p 1818
26. Goldberg A, Havelund K, McGann C (2005) Runtime verification for autonomous spacecraft software. In: 2005 IEEE aerospace conference. IEEE, pp 507–516
27. Graczyk R, Esteves-Verissimo P, Voelp M (2021) Sanctuary lost: a cyber-physical warfare in space. arXiv preprint arXiv:2110.05878
28. Grieves M (2014) Digital twin: manufacturing excellence through virtual factory replication. White Pap 1:1–7
29. Grieves M, Vickers J (2017) Digital twin: mitigating unpredictable, undesirable emergent behavior in complex systems. In: Transdisciplinary perspectives on complex systems. Springer, pp 85–113
30. Hartmann D, Van der Auweraer H (2021) Digital twins. In: Progress in industrial mathematics: success stories. Springer, pp 3–17
31. Hidalgo C, Vaca M, Nowak M, Frölich P, Reed M, Al-Naday M, Mpatziakas A, Protogerou A, Drosou A, Tzovaras D (2022) Detection, control and mitigation system for secure vehicular communication. Veh Commun 34. https://doi.org/10.1016/j.vehcom.2021.100425
32. Hou Z, Li Q, Foo E, Song J, Souza P (2022) A digital twin runtime verification framework for protecting satellites systems from cyber attacks
33. Hu Y, Zhu P, Xun P, Liu B, Kang W, Xiong Y, Shi W (2021) CPMTD: cyber-physical moving target defense for hardening the security of power system against false data injected attack. Comput Secur 111. https://doi.org/10.1016/j.cose.2021.102465
34. Jiang Y, Yin S, Li K, Luo H, Kaynak O (2021) Industrial applications of digital twins. Philos Trans R Soc A 379(2207):20200360
35. Joseph K, Sharma AK, van Staden R (2022) Development of an intelligent urban water network system. Water 14(9):1320. https://doi.org/10.3390/w14091320
36. Kang S, Chun I, Kim HS (2019) Design and implementation of runtime verification framework for cyber-physical production systems. J Eng 2019
37. Kaur MJ, Mishra VP, Maheshwari P (2020) The convergence of digital twin, IoT, and machine learning: transforming data into action. In: Digital twin technologies and smart cities. Springer, pp 3–17
38. Kim YV (2020) Satellite control system: part I—architecture and main components. In: Satellite systems—design, modeling, simulation and analysis
39. Kritzinger W, Karner M, Traar G, Henjes J, Sihn W (2018) Digital twin in manufacturing: a categorical literature review and classification. IFAC-PapersOnLine 51(11):1016–1022
40. Liu Z, Shi G, Meng X, Sun Z (2022) Intelligent control of building operation and maintenance processes based on global navigation satellite system and digital twins. Remote Sens 14(6):1387
41. Löcklin A, Müller M, Jung T, Jazdi N, White D, Weyrich M (2020) Digital twin for verification and validation of industrial automation systems—a survey. In: 2020 25th IEEE international conference on emerging technologies and factory automation (ETFA), vol 1. IEEE, pp 851–858
42. Luppen ZA, Lee DY, Rozier KY (2021) A case study in formal specification and runtime verification of a CubeSat communications system. In: AIAA Scitech 2021 forum, p 0997
43. Maesschalck S, Giotsas V, Green B, Race N (2022) Don't get stung, cover your ICS in honey: how do honeypots fit within industrial control system security. Comput Secur 114. https://doi.org/10.1016/j.cose.2021.102598
44. Min Q, Lu Y, Liu Z, Su C, Wang B (2019) Machine learning based digital twin framework for production optimization in petrochemical industry. Int J Inf Manage 49:502–519
45. Nweke LO (2017) Using the CIA and AAA models to explain cybersecurity activities. PM World J 6(12):1–3

46. Pacheco J, Hariri S (2016) IoT security framework for smart cyber infrastructures. In: 2016 IEEE 1st international workshops on foundations and applications of self* systems (FAS*W). IEEE, pp 242–247
47. Pedro L (2021) Validation and verification of digital twins
48. Pylianidis C, Osinga S, Athanasiadis IN (2021) Introducing digital twins to agriculture. Comput Electron Agric 184:105942
49. Rahiminejad A, Ghafouri M, Atallah R, Mohammadi A, Debbabi M (2022) A cyber-physical resilience-based survivability metric against topological cyberattacks. Institute of Electrical and Electronics Engineers Inc. https://doi.org/10.1109/ISGT50606.2022.9817513, https://www.scopus.com/inward/record.uri?eid=2-s2.0-85134892344&doi=10.1109%2fISGT50606.2022.9817513&partnerID=40&md5=1d64de0a25c495b8e072733ab8812dd7
50. Saratha SDC, Grimm C, Wawrzik F (2021) A digital twin with runtime-verification for industrial development-operation integration. In: 2021 IEEE international conference on engineering, technology and innovation (ICE/ITMC). IEEE, pp 1–9
51. Satam S, Satam P, Pacheco J, Hariri S (2022) Security framework for smart cyber infrastructure. Clust Comput 25(4):2767–2778. https://doi.org/10.1007/s10586-021-03482-2
52. Shangguan D, Chen L, Ding J (2020) A digital twin-based approach for the fault diagnosis and health monitoring of a complex satellite system. Symmetry 12(8):1307
53. Shao G, Jain S, Laroque C, Lee LH, Lendermann P, Rose O (2019) Digital twin for smart manufacturing: the simulation aspect. In: 2019 winter simulation conference (WSC). IEEE, pp 2085–2098
54. Sivalingam K, Sepulveda M, Spring M, Davies P (2018) A review and methodology development for remaining useful life prediction of offshore fixed and floating wind turbine power converter with digital twin technology perspective. In: 2018 2nd international conference on green energy and applications (ICGEA). IEEE, pp 197–204
55. Sleuters J, Li Y, Verriet J, Velikova M, Doornbos R (2019) A digital twin method for automated behavior analysis of large-scale distributed IoT systems. In: 2019 14th annual conference system of systems engineering (SoSE). IEEE, pp 7–12
56. Spanghero M, Zhang K, Papadimitratos P (2020) Authenticated time for detecting GNSS attacks. In: Proceedings of the 33rd international technical meeting of the satellite division of the institute of navigation (ION GNSS+ 2020), pp 3826–3834
57. Suo D, Moore J, Boesch M, Post K, Sarma S (2022) Location-based schemes for mitigating cyber threats on connected and automated vehicles: a survey and design framework. IEEE Trans Intell Transp Syst 23(4):2919–2937. https://doi.org/10.1109/TITS.2020.3038755
58. Tao F, Zhang H, Liu A, Nee AY (2018) Digital twin in industry: state-of-the-art. IEEE Trans Ind Inform 15(4):2405–2415
59. Verriet J, Buit L, Doornbos R, Huijbrechts B, Sevo K, Sleuters J, Verberkt M (2019) Virtual prototyping of large-scale IoT control systems using domain-specific languages. In: MODELSWARD, pp 229–239
60. Villalonga A, Negri E, Biscardo G, Castano F, Haber RE, Fumagalli L, Macchi M (2021) A decision-making framework for dynamic scheduling of cyber-physical production systems based on digital twins. Annu Rev Control 51:357–373
61. Vivek S, Conner H (2022) Urban road network vulnerability and resilience to large-scale attacks. Saf Sci 147. https://doi.org/10.1016/j.ssci.2021.105575, https://www.scopus.com/inward/record.uri?eid=2-s2.0-85121150901&doi=10.1016%2fj.ssci.2021.105575&partnerID=40&md5=951d1b2e5ea302f56e59d756f97ed744
62. Wilkinson DC, Daughtridge SC, Stone JL, Sauer HH, Darling P (1991) TDRS-1 single event upsets and the effect of the space environment. IEEE Trans Nucl Sci 38(6):1708–1712
63. Yang W, Zheng Y, Li S (2021) Application status and prospect of digital twin for on-orbit spacecraft. IEEE Access 9:106489–106500

Blockchain in Smart Grids: A Review of Recent Developments

Teng Yu, Fengji Luo, Quanwang Wu, and Gianluca Ranzi

Abstract As a trustworthy, decentralized ledging technology, blockchain has gained significant attention in the industry. In the last a few years, research and development work has been conducted in underpinning the capability of blockchain to support the secure, reliable and efficient operation of power and energy systems. This chapter provides a review for applications of blockchain in different operational aspects of smart grids. It introduces the basic concepts related to blockchain and smart grids and provides a taxonomy for blockchain-supported smart grid applications. It then reviews the recently representative work in each application area. The chapter is expected to provide a reference for researchers and engineers in the related fields.

Keywords Blockchain · Smart grid · Power system · Energy market · Distributed database

1 Introduction

Human society is facing the grand challenge of climate change. According to the 6th assessment report released by the Intergovernmental Panel on Climate Change (IPCC) on August 2021 [1], emissions of greenhouse gases from human activities are responsible for about 1.1 °C of warming since 1850–1900, and the report estimates that over the next 20 years, global temperature is expected to reach or exceed 1.5 °C of warming. One of the direct causes of climate change is the overuse of fossil fuels (typically oil, coal, and natural gas), which release Carbon Dioxide (CO_2) in their burning processes.

The ever-severe climate change challenge imposes an urgent demand to transforming the way we produce and consume energy. This indicates energy systems,

T. Yu · F. Luo (✉) · G. Ranzi
School of Civil Engineering, The University of Sydney, Sydney, NSW 2006, Australia
e-mail: fengji.luo@sydney.edu.au

Q. Wu
School of Computer Science, Chongqing University, Chongqing 644000, China

S. Pal et al. (eds.), *Emerging Smart Technologies for Critical Infrastructure*,
Smart Sensors, Measurement and Instrumentation 44,
https://doi.org/10.1007/978-3-031-29845-5_2

which have been powering the human society for centuries and traditionally rely on fossil fuels to generate energy, need to be transformed to operate towards a more sustainable manner. Driven by this, the concept of "smart grid" was proposed in the early twenty-first century and has been experienced worldwide development [2]. A representative feature of smart grids is the deployment of Renewable Energy Sources (RESs) directly deployed in energy customer side, e.g., Photovoltaic (PV) solar panels installed in buildings' façade and small wind turbines installed in urban areas. These geographically dispersed RESs in low-voltage power distribution systems are usually referred to as Distributed Renewable Energy Sources (DRESs) [3]. The increasingly widespread deployment of DRESs has been transforming the operation pattern of electrical power grids from a traditionally vertical to a distributed way. In the vertical pattern, electricity is generated from distantly large power plants, transmitted by high-voltage power transmission lines, and delivered to energy customers by low-voltage substations and feeders. In the distributed pattern, energy generated from DRESs can directly supply the on-site or nearby energy loads or it can be fed back to the grid.

The distribution feature of smart grids is also enabled by the deep integration of advanced Information and Communications Technology (ICT) [4], which assign automatic computing and communication capabilities to energy entities in the grid. For example: the Advanced Metering Infrastructure [5] enables energy loads to bidirectional communicate with the grid, e.g., an energy load can report its power consumption to the grid and can receive real-time energy price information and load control signals from the grid; the Internet-of-Things and automation facilities make energy entities be able to have direct communications and information exchange with each other.

The distributed feature of smart grids' operation indicates that it is desirable to develop distributed, robust information infrastructure to work with other ICTs to provide fundamental support in accommodating the complex interactions among different energy entities in smart grids. Emerged in the last decade, a distributed ledger technology called blockchain [6] has the potential to take this role. A blockchain is composed of a number of blocks that store data and are chained chronologically and are duplicated in a group of networked nodes. Based on this chain structure for data blocks and a certain consensus mechanism among the nodes, a blockchain system can ensure that the data stored in it cannot be changed by other parties. Such decentralized and data immutable features make blockchain have been actively applied in many industrial and financial systems, including smart grids.

This chapter is devoted to providing a more comprehensive, update-to-date review for the integration of blockchain in smart grids. To make the readers who are not familiar with the related concepts can understand the discussion in this chapter, we provide an introduction for blockchain and smart grids and outlines a taxonomy for the application sub-areas of blockchain in smart grids (Sect. 2). Then, we review the state-of-the-art in each sub-area based on the literature collected from 2020 to April 2022 (Sects. 3–6). Together with other review articles in this field [7–14], this chapter is expected to provide a useful and motivational reference to students, researchers and engineers who are interested in this topic.

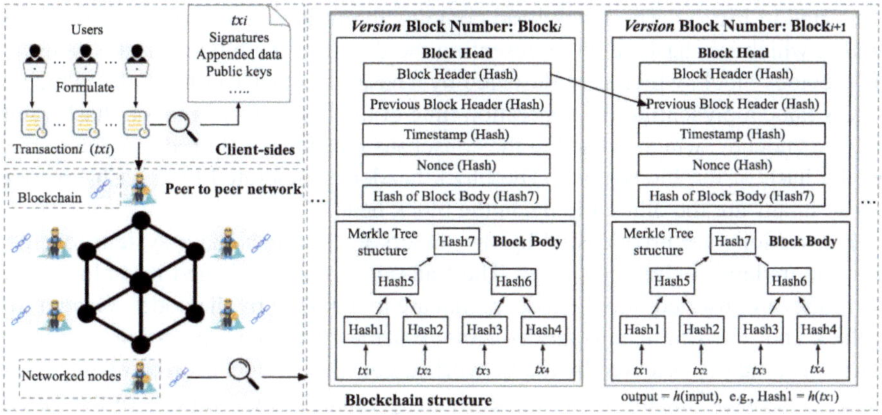

Fig. 1 Illustration of a blockchain

2 Introduction of Blockchain and Smart Grids

In this section, we will provide a brief and fundamental introduction to blockchain and smart grids. We will also outline a taxonomy about the current application scenarios of blockchain in smart grids.

2.1 Introduction of Blockchain

Blockchain was proposed by Nakamoto (it is believed this is a pseudonym) in 2008 [15]. In the paper, Satoshi describes a digital currency system called Bitcoin [15], which relies on a distributed ledging technology called blockchain to enable the participants to achieve secure payment without the intervention of the third party. Blockchain works in a P2P environment that consists of a number of networked nodes. Each node maintains a copy of a ledger (called a "blockchain") locally. The transactions generated in the system (i.e., the payment of Bitcoins) are packaged into blocks at a regular time interval, and the blocks are validated by the nodes and stored in the ledgers. The schematic of blockchain and its block structure is shown in Fig. 1.

A block consists of two parts: block body and block head. The data of multiple transactions are organized as a specific structure and are stored in the block body. In many blockchain systems, the transaction data is organized as a Merkle Tree [16] in a block body. Merkle Tree (also known as "hash tree") is a binary tree structure in which every leaf node is labelled with the hash value[1] of a data block, and every node that is not a leaf is labelled with the hash value of the labels of its child nodes (as

[1] A hash value is a numeric value of a fixed length that uniquely identifies data. Hash values are generated by hash functions, which map data of arbitrary size to fixed-size values (i.e., hash values) [17].

illustrated in Fig. 1). The block head contains the following information: (i) block header, which is a hash value representing the digest of the block; (ii) the digest of the previous block, which is represented by a hash value; (iii) a timestamp, which is a hash value representing the generation time of the block; (iv) a random hash value called a "nonce", which is used in the consensus mechanism (introduced later); and (v) the hash of the block body. The block header is generated by a hash function whose inputs are the other 4 items in the block head.

Transactions are broadcasted by the traders to the blockchain system together with the traders' digital signatures. The transactions are packaged into blocks and the blocks are inserted into the ledgers that are maintained by the nodes. Generating a block and inserting it into the ledgers are achieved through a certain consensus mechanism that is executed by the networked nodes in a decentralized manner. The word "consensus" means the nodes reach an agreement on the validity of the data in the block. There are different kinds of consensus mechanisms, such as the Proof-of-Work (PoW) mechanism [18], the Proof-of-stake mechanism [19], and the Byzantine Fault Tolerant mechanism [20].

The blockchain proposed in Nakamoto's paper is based on the PoW mechanism. In the PoW mechanism, the nodes compete to generate blocks to package the transactions generated in the system. Every time a node generates a candidate block, to obtain this right, it needs to find out a nonce value so that the hash of the candidate block satisfies a specific condition. This process is also called "mining" and the nodes are called "miners". The nodes set the nonce values in the head of the candidate blocks and broadcast them to the system. The candidate blocks will be checked by all the nodes in the system. If more than 50% of the nodes have verified a candidate block is valid (i.e., the nonce in the block head is the correct one), the candidate block will be regarded as a valid one. The first node that finds out the correct nonce will then receive certain rewards (in the Bitcoin system, the rewards are Bitcoins), and the candidate block generated by the node will be inserted into the blockchain and stored in the duplicated ledgers in the different nodes.

The transaction data stored in blockchain is traceable and immutable. Firstly, data traceability is ensured by blockchain's chain structure, because all the historical transaction records are stored in the blocks that are chained one-by-one. Secondly, since the blocks are chained by using their hash values, any change of the data in the body of a block will lead to the change of the block head's hash value, which will make all the subsequent blocks to be invalid. Unless the attacker can control more than 50% of the nodes to re-generate and re-insert all the subsequent blocks using the hash of the block containing the tampered data, the data that has been stored in the blockchain cannot be manipulated. When the number of nodes in a blockchain system is large, it is very difficult for an attacker to control more than 50% of the nodes; further, the mining mechanism in blockchain will cause big cost in block generation and will thus prevent a rationale attacker from regenerating blocks. Therefore, data immutability is ensured by blockchain.

Many blockchain implementations are integrated with smart contracts [21]. A smart contract is a container that stores data and executable codes. Execution of smart contracts is event driven: the codes in a smart contract are triggered to execute by certain pre-specified conditions. For example, in online commodity trading, a smart contract containing cash transfer codes will be triggered to transfer money from the buyer's digital wallet to the seller's wallet when the buyer confirms he/she has received the commodity.

2.2 *Introduction of Smart Grids*

Electrical power grids (also known as electrical power systems) are one of the most critical infrastructures in human society [22]. A power system generates electricity from power sources and transmits and delivers it to power the different entities that need electricity (e.g., buildings and homes, industrial equipment and public facilities). While power grids differ in scales and complexities, a typical power grid is with a 3-segment structure (Fig. 2), which consists of power generation, transmission, and distribution. Electricity is generated from the power plants in the power generation side; the electricity is sent to power transmission lines after its voltage is raised by transformers. Transmission lines transmit the electricity to energy load areas (e.g., cities), and the voltage is then lowered down by transformers there. The low-voltage electricity is delivered by feeders to individual energy loads (e.g., households, industrial equipment and public facilities).

Traditional power systems can hardly adapt to the changing environment caused by climate change and energy shortage. They heavily rely on fossil fuels (i.e., electricity is mostly produced by coal- or gas-fuelled generators), which leads to a huge amount of carbon emission. In addition, the communication flow between the energy load and the grid is unidirectional, which means the power generation is completely driven by the demand—the power utility company collects the power demand information of the loads and schedules power generation to satisfy the demand. In other words, the utility company cannot alter the power and energy consumption pattern of the load, but just generates electricity to satisfy the demand. This would lead to low efficiency of the power system. For example: the power demand of a load area at some time points (e.g., in hot summer days) could be very high. Since the power infrastructure (transmission lines, transformers, feeders, etc.) has a maximum capacity, if

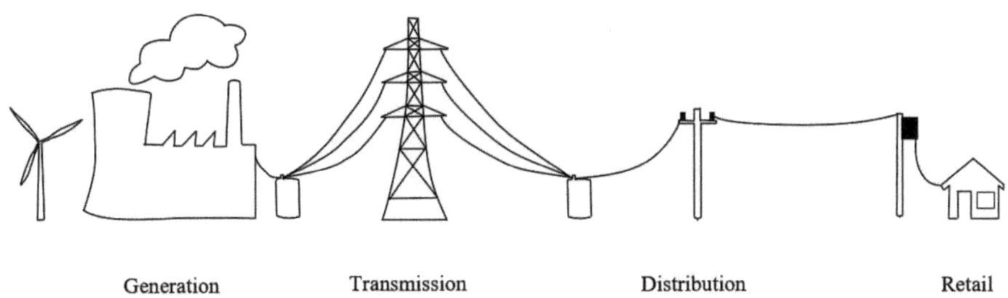

Fig. 2 Structure of a typical power system

the power demand exceeds the infrastructure's capacity, the utility company has to upgrade or shed the electricity supply to some loads. The former is costly, and the latter will result in blackouts and cause significant disturbance to people.

The proposal and development of smart grids represent the transformation of traditional power systems towards a more efficient and sustainable manner. The conceptual schematic of a smart grid is illustrated in Fig. 3. Compared with traditional power systems, smart grids have the following distinguishing features:

(1) *High penetration of renewable energy.* In a smart grid, RESs are pervasively deployed, which are typically wind turbines, solar thermal collectors, and PV solar panels. Some of them are built in the form of large-capacity wind or solar farms and are directly connected with power transmission networks through transformers. Besides this, considerable capacity of renewable energy is deployed in load areas as small-capacity, distributed RES units (e.g., PV solar panels mounted in buildings' roofs). The high penetration of renewable energy can reduce the dependency of human society on fossil fuels; in particular, the pervasively distributed RESs make energy loads able to perform self-energy supply; this can reduce the energy loss caused by long-distance electricity transmission and hence can drive the whole energy system to operate more efficiently and sustainable.
(2) *Bidirectional communication between energy loads and the utility company.* Compared with traditional power systems, Smart grids are featured by the

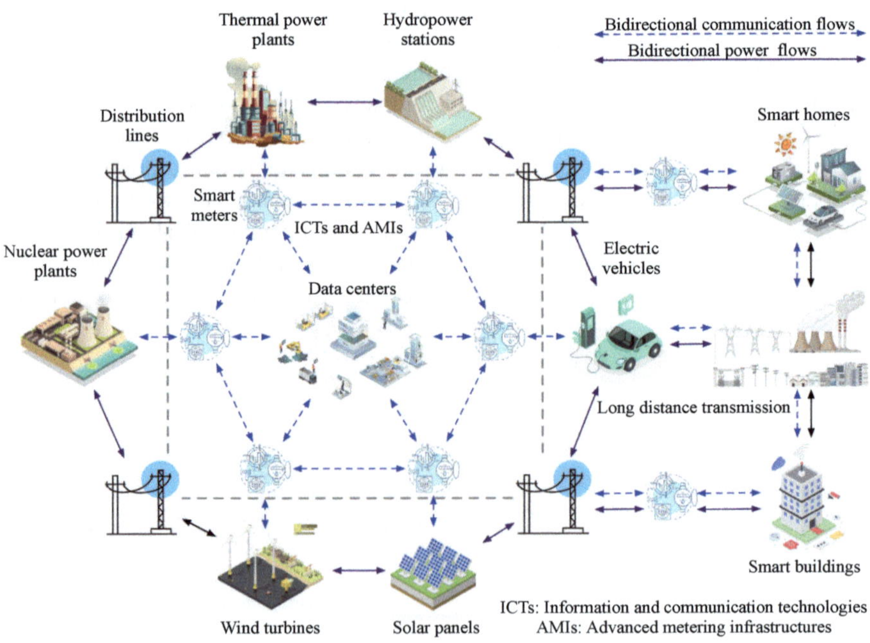

Fig. 3 Conceptual schematic of a smart grid

widespread deployment of Advanced Metering Infrastructure (AMI), which is backboned by smart meters [23]. Equipped with microprocessors, smart meters are capable of automatically measuring energy production and consumption and performing communication. AMI transforms the communication between energy customers and the grid from the unidirectional to bidirectional pattern. That is, an energy customer cannot only automatically send its power consumption to the grid, it can also receive information from the utility company.

With such a two-way communication, the utility company can set up incentive or price signals to influence the energy consumption behaviour of energy customers. For example, the utility company can set up high electricity prices in early evening hours when many energy customers have large power consumption that leads to high peak power demand in the grid; by receiving the price information, energy customers would choose to shift the operation of some appliances and equipment from early to late evening hours when the electricity price is low. In this way, the peak power demand in the early evening hours can be reduced, which prevents the grid from being overloaded. In other words, with bidirectional communication, the utility company can not only schedule power generation to satisfy the customer's energy demand, but can also take measures to influence the customer's power consumption pattern. This is referred to as the *power demand side management*.

(3) *High deregulation.* Traditional power systems, especially the ones decades ago, operate in a monopolistic mode—national or regional power utility companies manage all the segments of electricity generation, transmission, and distribution. In smart grids, the energy assets are owned by different parties. Electricity is regarded as a commodity and is traded by different stakeholders in a smart grid. Typically, power generation companies, which own power generation sources, sell electricity to power distribution companies or electricity retailers. Such trading can be performed by signing long-term contracts or in marketplace-like systems that are called "wholesale power markets". Distribution companies or retailers then re-sell the electricity to the end customers with prices that are larger than the prices they buy electricity from the power generation companies. The power utility companies which own power transmission assets and provide the electricity transmission service charge the power generation and distribution companies and retailers the transmission fee (i.e., the fee of using the power transmission infrastructure). The interested readers can refer to Refs. [24, 25] for more details detailed discussion about the deregulated power markets.

The increasingly prevalence of distributed RESs has been transforming many energy customers from pure energy consumers to energy "prosumers (producers and consumers)", meaning that they are capable of both generating and consuming energy. As a result of this transformation, energy trading now can also be performed among prosumers. For example, a household can sell the surplus energy generated from its rooftop PV solar panel to another household. Such a scenario is referred to as the peer-to-peer (P2P) energy trading and has been actively studied in recent years [1, 26].

(4) *Deep integration of advanced ICTs*. Compared with traditional power systems, smart grids deeply integrate with advanced ICTs. For example, the wide-area measurement devices (e.g., smart meters and phasor measurement units (PMUs)) makes it possible to real-time monitor the grid's operational condition in a fine-grained manner; Internet-of-Things and automation facilities enable different energy entities to communicate and exchange data with each other; advanced computing and control techniques provide support to develop sophisticated energy management strategies and make the grid be able to fast respond to disturbances and faults. All these drive smart grids to operate in a more intelligent, automatic, efficient, and fault-tolerant manner.

2.3 Taxonomy of Blockchain-Based Applications in Smart Grids

Essentially, blockchain is a decentralized database system that can ensure the traceability and immutability. With integration of smart contracts, blockchain systems have further evolved to an automatic task execution machine. These features indicate there is big potential for blockchain to provide multi-fold support to smart grids: firstly, it can provide a decentralized data management service to different energy entities in smart grids. The data generated from different sources in the grid can be stored and synchronized at different nodes, which can reduce the risk of single-point-of-failure; without the intervention of central management systems, the nodes in a blockchain cooperate through consensus mechanism to ensure the data stored in a blockchain cannot be tampered. This will improve data security and integrity in smart grids. Secondly, just as many successful blockchain-based applications in financial domain, blockchain has potential to fundamentally transform the way of trading energy and energy assets in smart grids towards a decentralized and trusted manner. This is particularly meaningful for accommodating the P2P energy trading requirement of small energy prosumers in low-voltage distribution networks that equip with renewable energy sources. Thirdly, with integration of smart contracts, blockchain can provide a trusted task execution environment on top of its trusted ledging and data management mechanism. This can help to build up application-oriented workflows in smart grids and facilitate their automatic execution.

We collect the literature published between 2020 and 2022 that study the application of blockchain in smart grids. In general, the application scenarios studied in the literature can be categorized into four classes: (1) energy and carbon credit trading, which studies using blockchain to support trading of electricity and carbon credits in smart grids; (2) smart grid data management, which studies using blockchain to manage the data in smart grids; (3) demand side management, which studies using blockchain to manage demand side energy resources and facilitate their demand response; and (4) grid operation and control, which studies using blockchain to support the smart grid's operation. The research in each scenario can be further

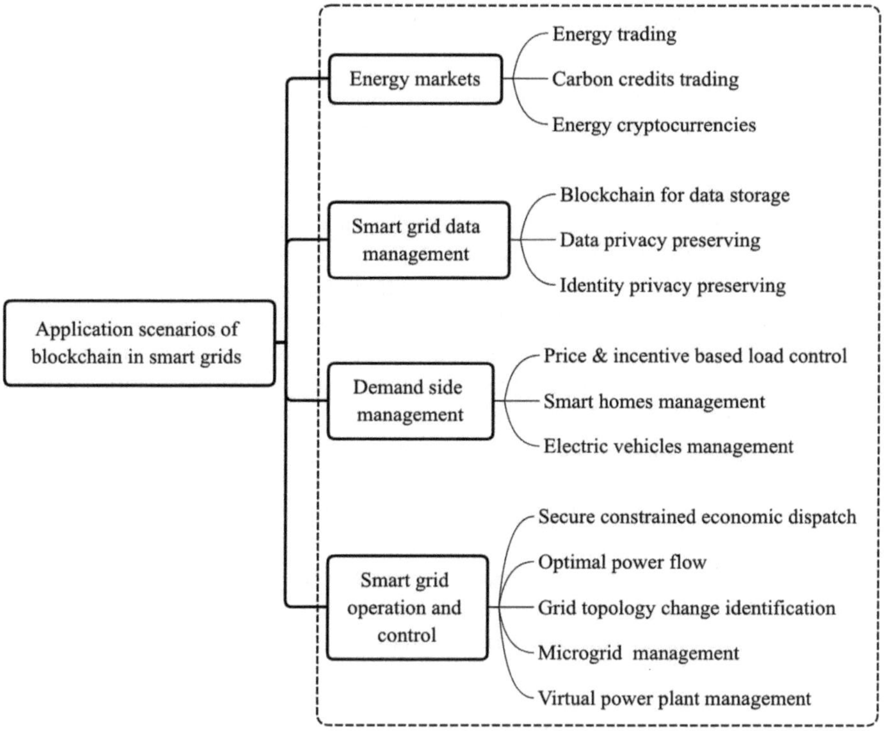

Fig. 4 Taxonomy of the application scenarios of blockchain in smart grids

divided into several specific topics as shown in Fig. 4. In the rest of this chapter, we provide reviews for each application scenario.

3 Blockchain-Supported Energy and Carbon Credit Trading

A direct application of blockchain is to support trading activities. In smart grids, these mainly involve energy trading and carbon credit trading. Besides, blockchain-based cryptocurrency has been developed to facilitate trading of energy and energy-related assets. The research in these aspects are reviewed in this section.

3.1 Energy Trading

In power systems, energy trading is typically performed by power generation companies and power distribution companies (or power retailers) in regional- or national-scale power markets [11]. The power system operator manages the market and settles the energy trading transactions. Since from the last a few years, DRESs have been being increasingly deployed in urban systems, and a considerable capacity of DRESs are deployed directly in end energy users' sites (e.g., in the form of rooftop or façade PV solar panels installed in buildings). The DRES deployment has naturally driven the emergence of a new energy trading paradigm, in which end users trade energy with each other. For example, a household can sell the surplus solar energy generated from its rooftop solar panels to its neighbors. Such an energy trading scenario is referred to as "P2P energy trading".

Since the participants in P2P energy trading are mostly prosumers that are with equal roles, blockchain can be well used to support this scenario. In general, there are two steps in the energy trading process between two traders: form an energy transaction and settle the transaction. The former refers to as the process in which the traders reach an agreement of energy trading price and amount, and the latter refers to as the process of carrying out the actual energy delivery and cash transfer. The workflow of blockchain-based energy trading is shown in Fig. 5.

3.1.1 Formation of Energy Trading Transactions

The formation of energy trading transactions mainly involves determination of the trading price and anergy amount. In some situations, the power network's physical constraints (such as power flow and nodal voltage constraints) also need to be considered in this determination process. The task of forming energy trading transactions can be compute-intensive, e.g., through solving a power network-constrained optimization problem.

Motivated by this, Refs. [27–31] develop blockchain systems new consensus mechanisms, which treat energy trading transaction formation tasks as the construction of consensus. Compared with the PoW mechanism in the classical blockchain (i.e., the Bitcoin system, where the calculation effort of a valid nonce is considered as the proof-of-work), in these new consensus mechanisms, the nodes of the blockchain consume considerable computing power and electricity not to just search for the valid nonce value, but to perform more "meaningful" tasks from the energy system's prospective, i.e., determination of energy trading prices and amounts. However, these transaction formation-based consensus mechanisms would lead to low operation efficiency and low throughput for the blockchain system, because every node is required to perform such compute-intensive transaction formation tasks. To address this issue, Refs. [32–35] propose a solution, in which the process of determining the energy trading prices and amounts are performed centrally and in an off-blockchain manner,

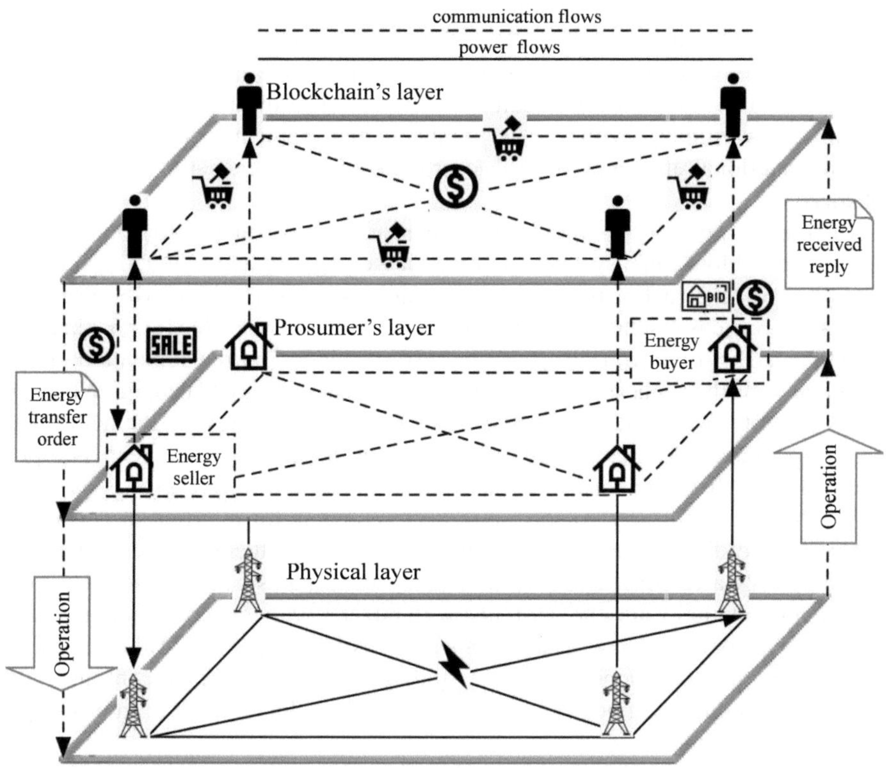

Fig. 5 Schematic of blockchain-based energy trading

and only the result (i.e., the formed transactions) is recorded in the blockchain. Nevertheless, this strategy is with a major drawback: since the transaction formation process is performed centrally and cannot be verified by the blockchain nodes, the fairness and trustworthy of the transaction formation result cannot be ensured. Four kinds of strategies are developed in the literature to ensure the fairness of energy trading transaction formation while maintaining the efficient operation of the blockchain system: (i) develop computational efficient transaction formation algorithms; (ii) use private or consortium blockchains; (iii) design new consensus mechanisms; and (iv) adopt multi-blockchain designs.

References [36–39] propose lightweight transaction formation algorithms that can be executed by smart contracts in the blockchain. Such algorithms only use a small number of computational iterations to determine the energy trading price; therefore, they are suitable for being deployed in smart contracts and fast executed without significantly affecting the blockchain system's throughput. However, some energy trading applications involve sophisticated and complex negotiation among the traders and market and market clearing logics; in these situations, it would be difficult to develop lightweight transaction formation algorithms. References [31,

37–45] utilize private or consortium blockchains to achieve the trustable transaction formation in P2P energy trading. Such types of blockchains are not managed by a large number of nodes that can freely join or leave the blockchain, but only a few dedicated nodes that are configured with powerful hardware, i.e., CPU cores and random-access memory. With reduced number of management nodes, the throughput and computational efficiency of private or consortium blockchains can be greatly enhanced.

References [29, 46] propose new consensus mechanisms to improve the throughput of blockchain systems in processing energy trading transactions. Reference [29] proposes a Practical Byzantine Fault Tolerant (PBFT)-based Proof-of-Reputation consensus scheme, in which the node who is responsible for bookkeeping is not selected by solving puzzles, but is based on the nodes' current reputation values. The node with the highest reputation will be selected for bookkeeping. When the leader generates a new block and broadcasts it to other nodes, only the nodes with top-N reputation values are eligible for verifying the new block (here N is a pre-specified constant). The consensus mechanism developed in Ref. [46] is based on PoW—the nodes still need to solve puzzles to compete for the right of generating blocks for recording the energy trading transactions. Nevertheless, different with the classical PoW-based consensus mechanism, in the mechanism in Ref. [46], the nodes with high reputation values are assigned with puzzles with less computational difficulty, and they therefore have a large chance to win the competition. This encourages the nodes that perform honest in energy trading to participate in the consensus and consume less computing power to solve the puzzles, which therefore improves the system's throughput.

Reference [47] designs a dual blockchain-based system to determine the energy trading price. The system is backboned by two blockchains that interact with each other to process energy trading transactions. Based on the dual blockchain structure, a compute-intensive energy trading pricing algorithm is decomposed into two parts of executable codes. One part is stored in a smart contract that is located in one blockchain and the other part is stored in a smart contract in the other blockchain. The pricing task is then carried out through an iterative interaction process between the two smart contracts: one smart contract generates an energy trading price and sends it to the other smart contract, and the latter perform adjustment on the price and sends it back to the first smart contract. This process terminates when the second smart contract does not generate any further adjustments to the price. With this mechanism, the workload of executing the compute-intensive pricing algorithm is distributed into two separate blockchains that can be managed by different nodes, which can avoid computational overloading for a single blockchain.

3.1.2 Settlement of Energy Transactions

In a typical process of transaction settlement, an energy seller transfers the agreed amount of energy to an energy buyer, and the smart meter of the seller generates an Energy Transfer Record (ETR) that records the energy the seller has actually transferred to the buyer. The ETR is then sent to the buyer, and the buyer checks whether it has received the agreed amount of energy based on the energy received record (ERR) generated from its smart meter. If yes, the buyer transfers cash to the seller according to the agreed energy trading price.

References [46, 48, 49] designs new consensus mechanism for achieving secure and trustable settlement of energy trading transactions. These consensus mechanisms share the same logic as follows: the formed energy trading transactions are firstly stored in a blockchain; then, after the blockchain has received ERRs and ETRs, the consensus process is initiated, in which the nodes verify if the data recorded in the ERRs and ETRs matches the transactions or not. The transactions that are considered as valid in the consensus then will be settled. In these mechanisms, when performing the verification, the nodes need to retrieve the energy trading transactions that have been stored in the previous blocks of the blockchain, which imposes additional data retrieval cost to the system. To reduce this cost, Ref. [32] sets up a central node, which is responsible for collecting the energy trading information from the participants (ETRs, ERRs, etc.). The node then sends all the information as a batch to the blockchain for verification. In this way, the nodes of the blockchain can perform transaction settlement verification directly on the data sent from the central node and does not need to retrieval the historical blocks.

Smart contracts are also used to facilitate the settlement of energy trading transactions [33–35, 50–56]. The settlement conditions and logics are programmed into smart contracts as executable codes. Once the conditions of transaction settlement have been satisfied, the energy traders send their ERRs and ETRs to trigger the smart contracts, which automatically execute the transaction settlement.

3.1.3 Incentive and Penalty Mechanisms for Energy Trading

One important concern in P2P energy trading is to encourage the traders to honestly and correctly perform energy trading and avoid misbehaviors or malicious actions. For example, once the agreed amount of energy has been delivered by the seller, the buyer should transfer the payment to the seller in a timely manner. References [57, 58] develop incentive and penalty mechanisms for this purpose, and the incentive and penalty logics are deployed in smart contracts. In the blockchain system developed in Ref. [57], a reputation score is assigned to every energy trader, which is determined based on the trader's historical energy trading records. For example, it a trader always timely transferred the correct amount of money to the sellers in the past when it purchased energy, the trader will be assigned a high reputation score. Based on the reputation scores, an energy trader can decide from whom it wants to purchase energy or to whom it intends to sell energy. In the penalty mechanism in Ref. [58], only the energy trading behaviors of the energy traders in the last time are considered. For instance, if an energy seller did not transfer the agreed amount of energy to an energy buyer, such a misbehavior will be recorded in the blockchain, and the seller will be prohibited from participating in energy trading over a certain time period until the seller has paid a fine for its misbehavior.

References [59–61] propose deposit schemes for energy trading in the blockchain environment. In the schemes, the energy traders need to deposit a certain number of tokens in a smart contract in advance of trading energy. Then, if the system detects misbehaviors from a trader (e.g., it does not transfer money to a seller from which it buys energy), the deposit of the trader will be forfeited by the smart contract.

3.2 *Carbon Credit Trading*

As a countermeasure for climate change, the Carbon Credit (CC) market has gained active practice and discussion in the last a few years. Carbon credits began as part of the 1997 UN Kyoto Protocol, the first international agreement to cut CO_2 emissions [12]; afterwards, CC market projects was launched in many countries such as China and Australia. The initiative of CC markets is to promote the carbon emission reduction in the society and the utilization of renewable energy. A CC is a permit that allows the owner to emit a certain amount of CO_2 or other greenhouse gases. A CC market is a marketplace in which the CC owners (i.e., the entities which are involved in carbon emission) can sell or buy CCs with each other. As pointed out in Refs. [12, 62], there are three basic requirements in the establishment of CC markets: (i) trustable carbon emission monitoring, verification and reporting (MRV); (ii) secure and efficient CC trading; and (iii) sufficient incentive for encouraging the carbon emission-related entities to participate in. Blockchain has shown its great potentials to support the above basic requirements and the schematic of the blockchain-based carbon credit trading is presented in Fig. 6.

Just as energy trading, blockchain can naturally be utilized for supporting CC trading. Reference [63] develops a blockchain-based system for achieving trustable carbon emission MRV in a CC market environment. In the system, the smart meters of the traders of the CC market send the traders' carbon emission information to smart contracts; then, the smart contracts monitor the traders' carbon emission amounts and send warnings to the traders who have high emission levels. References [63–70] develop blockchain-based CC markets. Similar to blockchain-based energy trading,

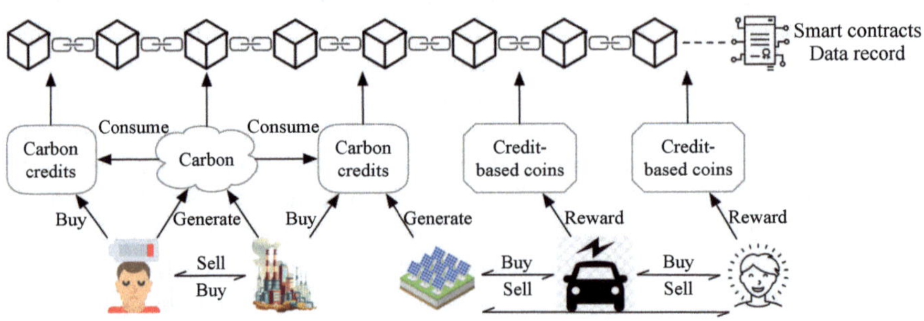

Fig. 6 The schematic of the blockchain-based carbon credit trading

in these CC markets, smart contracts are used to store the CC traders' bids, and the CC transactions are verified based on the consensus mechanisms and are stored in the blockchains.

Just as fiat-currencies, CCs can be traded by cross-region or even cross-nation entities. This imposes the requirement of high throughput to the underlying blockchain system. Reference [63] describes a conceptual hierarchical blockchain prototype for supporting cross-border CC trading. In the higher level, a public blockchain is set up for processing the cross-border CC trading transactions, and it can be managed by the operators in different countries. In the lower level, multiple local blockchains are setup, where each one processes the CC trading transaction within a certain region/country and is managed by a local operator.

Reference [71] proposes a blockchain-based platform for accommodating energy and CC trading. The system hosts smart contracts containing incentive logics, which encourage the energy entities to invest in solar panels. An entity that has installed solar panels will not only be able to obtain economic benefit via selling the solar energy to others, but will also be rewarded by CCs. Reference [70] proposes a blockchain system for managing the CCs of different energy entities in the P2P energy trading process. In the system, a penalty mechanism is deployed into a smart contract. If an energy trader's remaining CC amount is lower than a pre-specified threshold, the smart contract will be triggered to execute the penalize logic and the trader will be penalized.

3.3 Energy Cryptocurrencies

Cryptocurrency refers to as a form of currency that exists digitally and uses cryptography to achieve secure transactions. Just as the famous cryptocurrency Bitcoin, which is the first application blockchain, other cryptocurrency systems can be developed in blockchain to support trading of energy and energy-related assets. In general, these blockchain-enabled cryptocurrencies can be categorized into fiat currency-based cryptocurrencies [72] and original cryptocurrencies [73]. The former refers to as a type of digital currency that is bonded to fiat currencies, and the latter refers to as the digital currencies that only circulate in specific virtual environments and do not bond to any fiat currencies.

3.3.1 Fiat Currency-Based Cryptocurrencies

Reference [74] advocates to use blockchain-enabled, fiat currency-based cryptocurrencies as the payment method for energy trading. The literature lists two advantages of fiat currency-based cryptocurrencies: (i) less risk of monetary fluctuation, and (ii) reduced policy risk, considering the situation that in some countries, the original cryptocurrencies are restricted to be used.

On this basis, Refs. [27, 33, 38] develop fiat-currency cryptocurrencies for energy trading based on the fiat currencies in different countries. For example, Ref. [33] develops the Euro-Token cryptocurrency for P2P energy trading in a blockchain environment, where the tokens can be exchanged with Euros at a 1:1 rate. Reference [75] proposes a digital coin-based loan scheme to support energy trading. In the scheme, a digital credit bank is set up for issuing fiat currency-based digital coins. If an energy buyer does not have enough coins, the bank can lend coins to the buyer based on the buyer's credit value that is stored in a blockchain.

3.3.2 Original Cryptocurrencies

Blockchain-enabled original cryptocurrencies (such as Bitcoin) are incentive in nature. Similar to stocks, the values of these original cryptocurrencies can fluctuate, and this incentivizes the nodes of a blockchain to contribute more effort in maintaining and developing the blockchain. Reference [76] develops a blockchain-based energy trading system called Energy Web, in which the traders use an original cryptocurrency called Energy Web Token to trade energy. The tokens are awarded to the nodes which make contributions to secure the blockchain, such as verifying the transactions and generating blocks. Reference [27] designs two types of currencies for energy trading in blockchain systems. One currency is fiat currency-based, which is used for exchange rate stability; another is an original currency, which is released by the blockchain to incentivize the blockchain nodes to manage the system.

The incentive mechanism of original cryptocurrencies can also be associated with the energy-efficient actions the users have taken. Reference [77] develops a blockchain-based cryptocurrency called CERCoin. In the system, the users earn coins based on their "green mileage" values. The green mileage measures the distance a user has driven using renewable energy to power his/her EV. Reference [78] designs a blockchain-enabled original cryptocurrency and its whole release is based on carbon credits. In the system, a local government issues a certain number of tokens in a blockchain, and each taken is associated with a certain amount of CO_2. If an energy consumer plan to purchase electricity that is generated from non-renewable sources, the consumer is required to pay tokens to compensate for this environmentally unfriendly energy purchase. To prevent the opportunistic carbon hoarding behaviour,[2] Refs. [67, 79] set an expiry date for the carbon credit-based token. This restricts that if the taken is not spent by the owner before the expiry date, it will be regarded as invalid.

The above original cryptocurrencies belong to the so called "fungible tokens", meaning that the tokens (or known as digital coins) in a cryptocurrency system are fully interchangeable, identical and divisible. Recent research efforts [80–82] study the design of Non-Fungible Tokens (NFTs) [83] for energy trading. Unlike fungible tokens, NFT is essentially a data unit generated by blockchain systems. The data units

[2] Hoarding behaviour means the token owners deliberately hoard the tokens for some specific purposes.

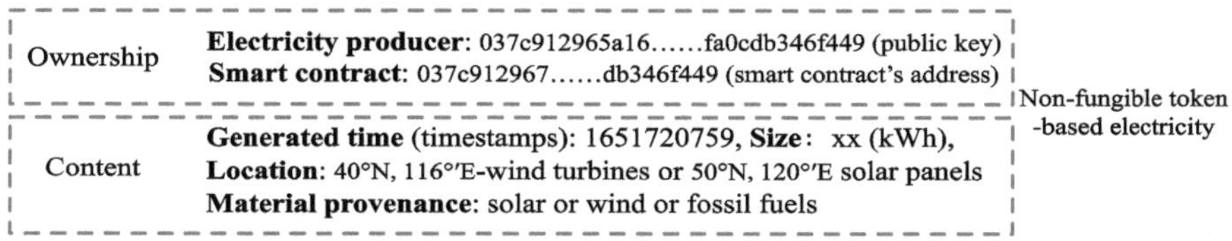

Fig. 7 A sample of non-fungible token-based electricity

are different with each other, making NFTs non-interchangeable and indivisible. Reference [82] uses NFTs, which are generated by smart contracts in a blockchain, to uniquely label the electricity generated in different time, with different amounts, at different locations, and from different sources. Such an NFT-based electricity labelling scheme is illustrated in Fig. 7.

4 Blockchain-Enabled Data Management in Smart Grids

Smart grids are featured by the integration of pervasive sensing and measurement devices (e.g., smart meters, phasor measurement units and IoT sensors). This results in the vast volume of data generated in the grid. For example, it is estimated that over 100 GB data can be generated in a day by 100 PMUs [84]. Storing and managing the data generated from pervasive sensing and measurement devices in smart grids are challenging tasks. One solution is to upload the data to cloud systems [85]. Nevertheless, managing such a tremendous volume of data in data centers is still a non-trivial challenge. Although blockchain is originally designed as a ledger for recording the transactions with digital coins, research efforts have been conducted to exploit the possibility of using blockchain as a kind of distributed database to store the data in smart grids [86–94].

4.1 Use Blockchain as a Distributed Database in Smart Grid Data Management

References [86–88] conduct preliminary research on using blockchain as a distributed database for storing data generated by smart meters. The smart meter data is stored and synchronized in different nodes, where each node maintains a blockchain ledger. Relying on blockchain's principles, the data can be prevented from being tampering as long as the ledgers in no more than 50% nodes in the system are maliciously modified; also, the consensus mechanism of blockchain makes the validity and correctness of the smart meter data can be checked collaboratively by the nodes before it is recorded into the blockchain. Taking the advantage of data traceability provided by blockchain, Ref. [89] develops an energy tracing system, which tracks the source of the energy consumed by the customer. In the system, a blockchain is used as a database for storing two aspects of information: (i) the identification of the energy supply source of the customer; and (ii) the type of the energy source (e.g., a fossil-field generator or a wind turbine). In this way, when necessary, the source and type of the energy used by the customer can be traced through the blockchain. For example, when the customer participates in a carbon credit program, the other stakeholders can use the system to trace the source of the energy used by the customer and make decision on the amount of carbon credits granted to the customer.

In a blockchain system, the data is duplicated in different nodes. Although such a redundant data storage mechanism plays a fundamental role for ensuring the data immutability in blockchains, it is associated with non-trivial data storage and management cost. References [90–94] develop solutions to reduce the data storage and management costs of using blockchain to store data in smart grids. The solution in Ref. [90] is based on the concept of private blockchain. The authors develop a dual-blockchain system, which consists of a private blockchain and a consortium blockchain. The private blockchain stores the data from DERs and the consortium blockchain processes the energy trading transactions. The private blockchain can provide a secure environment for data management since it is managed by authorized nodes. Also, compared with public blockchains, the private blockchain consumes less memory for data storage.

References [91–94] provide an alternative solution for reducing the cost due to the redundant data storage in blockchains. The systems developed in combines the InterPlanetary File System (IPFS) [95, 96] and blockchain to store energy data. IPFS is a distributed data storage system that enables end users to offer devices for data storage purpose shown as Fig. 8. By integrating IPFS and blockchain [91–94], the raw data to be store is firstly divided into multiple parts, and each part is hashed. A Distributed Hash Table (DHT) is set up, which records the head hash (i.e., the hash of the raw data), the hash values of the divided parts, and the identifications of the data storage devices. Only the head hash is stored in a blockchain, while the actual data is stored in the IPFS in a distributed manner. Through accessing the head hash in the blockchain and associating the head hash with the information of the different parts of the data, an external data user can locate and access the actual data stored in IPFS.

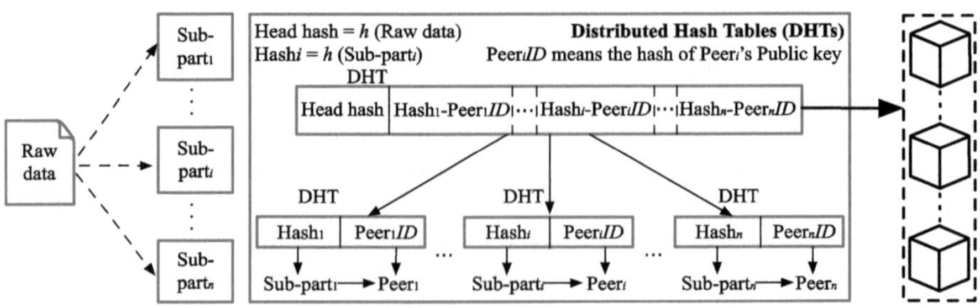

Fig. 8 The schematic of IPFS-based data storage structure

While blockchain's basic working mechanism ensures data immutability and traceability, it does not ensure data privacy—this is an important concern when using blockchain to store data in smart grids. References [7, 97–103] integrate additional techniques into blockchain to achieve *data privacy* and *identity privacy* in smart grids. The former refers to preserving privacy of the data collected by sensors and measurement devices, and the latter refers to preserving privacy of the location and identity of the entities in smart grids.

4.2 Data Privacy Preserving in Blockchain-Based Smart Grid Data Management

In the literature, data privacy preserving in blockchain-based smart grid data management systems is achieved by using the Asymmetric Cryptographic Technique (ACT, also known as public-key cryptography) [104], which is a representative encryption method and has been widely used in blockchain. In ACT, any person can encrypt a message using the message receiver's public key, but that encrypted message can only be decrypted with the receiver's private key. Reference [7] develops an ACT-based method for preserving the privacy of energy trading data in a blockchain environment. In the method, a user (denoted as user A) uses the public key of another user (denoted as user B) to encrypt its power consumption data and sends the encrypted data to a blockchain. The encrypted data can only be decrypted by user B using its own private key.

Although ACT can prevent data from being exposed to unauthorized parities, operations performed on the data still need to be performed on the plaintext. For example, to calculate the total power load of an area, the encrypted smart meter readings of the different users need to be transmitted to an energy management system and be decrypted there, and then the summation operation needs to be performed on the decrypted smart meter readings to obtain the total load. In this process, a malicious attacker could eavesdrop the smart meter data when it is decrypted. To tackle this, Ref. [97] integrates the Paillier Homomorphic Encryption (PHE) [98] technique and blockchain to achieve secure smart meter data aggregation. PHE is an encryption technique which can enable the summation operation to be performed directly on the encrypted numbers without revealing the plaintexts of the numbers. In the system developed in Ref. [97], multiple aggregators are setup and each one uses PHE to calculate the total load of a certain area based on the encrypted smart meter data in that area; the regional load data is then sent to blockchains for storage.

The smart meter data aggregation scheme developed in Ref. [97] relies on aggregators that are essentially centralized entities, and this potentially imposes the single point of failure to the system. Reference [99] designs a fully decentralized, secure smart meter data aggregation scheme, which preserves data privacy based on the data segmentation, the Secure Multi-Party Computation (SMPC) technology [100], and blockchain. Data segmentation means randomly dividing the original data into multiple parts. SMPC is a privacy preserving-based computation paradigm that is usually supported by PHE; it enables multiple non-trusted parties to collaboratively perform a computation task without exposing each party's private data. In the scheme in Ref. [100], each smart meter splits its reading into two segments, and then sends one segment to other smart meters. Meanwhile, each smart meter receives data segments from other smart meters. Then, the smart meter calculates the sum of these data segments and sends the result to a blockchain for further aggregation.

Reference [101] develops a zero-knowledge proof (ZKP)-based scheme for storing smart meter data into blockchain. ZKPs [105–107] are a special form of interactive protocols between a so-called "prover" and a "verifier". In ZKPs, the prover can convince the verifier about a specific statement without revealing any additional information about the statement. In this scheme, an energy consumer firstly uses a ZKP scheme to generate a statement, which claims that the smart meter reading of the consumer has been updated. Then, the statement is uploaded to blockchain and the nodes in the blockchain verify whether the statement is true or not without knowing the actual value of the reading.

4.3 *Identity Privacy Preserving in Blockchain-Based Smart Grid Data Management*

In blockchain systems, a public key of a participant represents the participant's identity and it does not directly associate with the participant's real-world identity. However, in many application scenarios that involve monetary transactions (e.g., energy trading), there will be a link between a participant's public key and real-world assets (e.g., bank accounts). In this situation, attackers may infer the participant's real-world identity through analysing the relationship between the public key and the real-world assets.

Aiming at preserving privacy of the participants' real-world identities, Refs. [100, 102, 103] develop solutions to prevent the participants' public keys in blockchain systems from being exposed. In Ref. [100], a one-time address generation method is used for this purpose. With this method, a participant in the blockchain generates different public keys every time when it needs to encrypt the data. A public key is generated by using the participant's private key to sign a randomly generated binary value. This method can prevent the participant' real-world identity from being revealed from a fixed public key. References [102, 103] uses the permissioned blockchain to protect participant's identity. Unlike conventional blockchains in which anyone can participate, permissioned blockchains only allow the authenticated participants to access the blockchain system. In a permissioned blockchain, the participant's public keys are only exposed to the authenticated participants, which can hence prevent the participants' identity information from being eavesdropped by the external attacker.

5 Blockchain-Supported Demand Side Management

Smart grids are characterized by the active participation of end energy customers. Through demand response programs, customers can change and reshape their energy consumption patterns and assist the grid to achieve certain operational objectives, such as peak power reduction and voltage regulation. In recent years, research efforts have been conducted to use blockchain to facilitate demand side management.

References [108–111] aims at addressing some fundamental issues in demand side management: encourage energy customers to participate in demand response programs, facilitate the interaction between the customers and the grid, and verify and monitor the customers' execution of demand response services. Reference [108] proposes a blockchain-based incentive mechanism for this issue. The mechanism uses a proof-of-energy saving consensus algorithm to encourage customers to reduce their energy consumption in peak load hours. In the system, a node represents an energy customer. The nodes then compete for being chosen to perform bookkeeping, i.e., packaging the contracted and actual load reduction and the customers' reputation values into blocks and inserting the blocks into the blockchain. The chance of a node to be chosen is based on 3 factors: (i) the demand response service margin the node can provide to the grid; (ii) the node's reputation score, which depends on its historical behaviours in the system; and (iii) the node's compliance ratio factor, which represents the completion degree of the node in the historical demand response programs. The node that is chosen for bookkeeping will receive a certain amount of reward.

Reference [109] develops another blockchain-supported incentive scheme to encourage end customers to participate in demand side management programs. In the scheme, a node of the blockchain is a prosumer configured with flexible energy resources (e.g., shiftable appliances). The grid operator submits a request to the blockchain, which contains the information of: (i) power consumption re-shaping targets, and (ii) the reward the grid intends to pay to the customers for the load-reshaping service. In the meantime, each prosumer submits its power consumption data to a smart contract, which calculates the amount of shiftable load of the prosumer. Based on the grid's request and the shiftable load information of the prosumers, a central agent selects the prosumers to perform load shifting and determine the allocation plan of the reward provided by the grid among these prosumers.

Reference [110] proposes a blockchain-based demand response service tracking and data sharing system. The system works as follows: a customer firstly submits a transaction to a blockchain through the AMI, and the transaction is composed by the information of the customer's controllable loads as well as the customer's signature. Smart contracts in the blockchain are then triggered to verify the transactions: if it is valid, the smart contracts will send the customer an endorsement. The customer then forwards the endorsement to the service requestor (which can be a grid operator) together with the transaction. The service requestor aggregates all the valid transactions and makes load control plans for each customer. Afterwards, the service requestor package all the load control decisions into a transaction and sends it to the blockchain.

Reference [111] designs a blockchain-based system for enabling price-based demand response. In the system, the customers periodically upload their forecasted power demand to the blockchain; based on the forecasted load, the grid makes time-varying electricity prices and sends them to the blockchain. After receiving the prices, the customers make load shifting decisions and their actual power consumption data is uploaded to the blockchain. The data is then checked by a smart contract, which verifies if the customer has successfully completed the demand response requirement: if yes, the customer will be rewarded; otherwise, penalty will be applied. Through such a reward and penalty mechanism, the system can encourage honest customers to consistently engage in demand response programs while restricting malicious customers to participate.

The above literature addresses the fundamental demand side management problems, and the techniques are applicable to any types of energy customers. Apart from this, some literature focuses on utilizing blockchain to manage two of the most important energy customers in modern urban systems: smart homes and electric vehicles (EVs).

5.1 Blockchain-Based Solutions for Smart Home Management

The concept of smart home refers to as the integration of different services within a home, aiming to provide healthcare and comfortable living environment for inhabitants [112, 113]. In the smart grid context, smart homes can take the advantages of advanced ICTs to benefit both the occupants and the grid. Through automatically and intelligently control appliances, smart homes cannot only increase the occupants' comfort and save energy bills to the occupants, but also help the grid to improve the grid's operation (e.g., change the home power consumption to regulate the voltage on feeders).

Featured by deep penetration of IoT and building automation facilities, smart homes operate in a data-intensive manner. Nevertheless, the IT devices in household environment are usually with limited computation, bandwidth and data storage capabilities, making it challenging to effectively manage the data generated in smart homes. Possible solution for this includes: (i) upgrade the home computing and data storage power, or (ii) upload the data to a third party (e.g., a cloud) for central management. The former is associated with non-trivial investment and maintenance cost and the latter would result in authority and privacy issues.

Literature [20, 114–118] develop blockchain-based solutions for smart home management. While blockchain can be used as a secure data storage infrastructure, the conventional blockchain prototype (i.e., the one described in Nakamoto's paper [15]) can hardly store the large volume of data in a smart home in an efficient and economical manner. To enhance the data storage efficiency, Ref. [114] proposes a new blockchain system, in which the nodes are authenticated cloud servers that have powerful computing and data storage capabilities. The manager/occupant of a smart home can send the data to the blockchain and access and operate the data using his/her private key. Aiming at dealing with the limited communication capacity issue of smart meters, Ref. [115] proposes a blockchain-based, IoT-enabled smart home energy management system. In the system, smart meters not only participate in the smart homes' energy management, but also act as the blockchain's nodes to take the responsibilities of generating and verifying energy transactions. To improve communication efficiency, a modified PBFT consensus mechanism is design in to reduce the communication complexity in the consensus process. Through optimizing the prepare and commit phases in PBFT [20], the modified PBFT can reduce the communication complexity of the traditional PBFT from $O(n^2)$ to $O(n)$, making the system can be well adapted to home devices.

Reference [116] systematically studies the use of blockchain to accommodate the pervasive IoT devices in smart homes. The reference describes 3 types of data management schemes: an IoT-IoT scheme, an IoT-blockchain scheme, and a hybrid scheme (Fig. 9). In the IoT-IoT scheme, the smart home devices and other entities (e.g., home energy management systems) exchange data directly through the machine-to-machine communication channels among them, and only the home management information agreed by the devices/entities (e.g., turn on a smart light at a specific time point) are uploaded to a blockchain. In the IoT-blockchain scheme, all the data and information in a smart home are uploaded to a blockchain for secure storage and authorized access. In the hybrid scheme, the optimal data routing plan is determined among the smart home IoT devices. Based on the optimized data routing plan, the IoT devices decide which data should be directly exchanged among them through machine-to-machine communication channels and which data should be submitted to the blockchain. In this way, the hybrid sharing scheme can alleviate the limitations in the first two schemes, i.e., the privacy issue in the IoT-IoT scheme and the heavy storage burden in the IoT-blockchain scheme.

Direct load control (DLC) [117] is a typical demand side management approach, which refers to have an external entity (typically a power utility company) to remotely

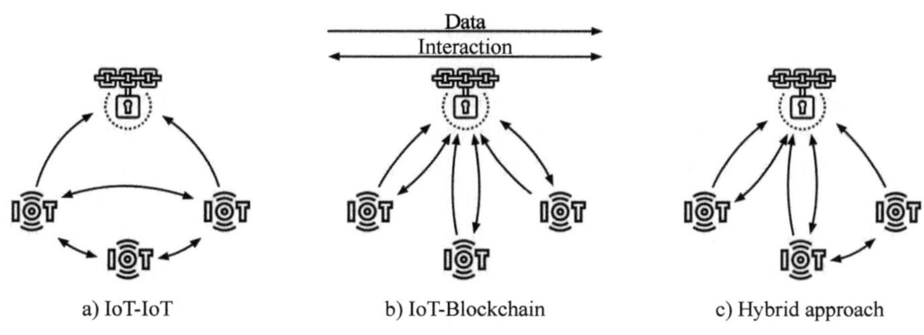

Fig. 9 Three types of data management schemes

control the customer-side appliances subjected to certain objectives, e.g., peak load reduction and feeder voltage regulation. Reference [118] designs a blockchain-based scheme for DLC. In the system, a smart home's gateway is managed by a blockchain. To launch a DLC operation, the external entity is firstly checked by a smart contract in the blockchain subjected to the access conditions that are pre-specified in the smart contracts. If the entity satisfies the conditions, it sends the load control agreement information between the entity and the home to a smart contract to request for verification. If the agreement information has passed the verification, the gateway will be opened to the external entity and the latter can perform remote load control to the appliances.

5.2 Blockchain-Based Solutions for EV Management

Besides acting as a transportation tool, EVs are also playing an increasingly important role in modern energy systems. By properly making charging/discharging actions to the batteries, plug-in EVs can provide demand response service to the grid. EVs can also be used as an energy backup to households. These scenarios are referred to as "Vehicle-to-Grid (V2G)" [119] and "Vehicle-to-Home (V2H)" [120] integration. Despite the merits of V2G/V2H, it is a non-trivial task for managing a large number of EVs in power distribution systems—this involves promptly monitoring the EVs' information (locations, the batteries' state-of-charge, and the EVs' health condition) and facilitating the verification and settlement of the demand response services provided by the EVs. In addition, it is also desirable to include the EV owners into the management loop; this would involve generating consultant information or recommendations to an EV owner in terms of efficiently using the EV.

Blockchain-based EV management solutions are developed in Refs. [29, 121–126] for the above purposes. Reference [121] proposes a blockchain-based EV battery management system. The nodes of the system estimate the remaining energy amounts of a population of geo-graphically distributed EVs; the estimation results are verified and stored in blockchain. An EV owner, who acts as a participant of the system, can access the data in the blockchain and make a battery swap plan based on the estimated remaining energy amount in his/her EV's battery.

References [122, 123] study using blockchains to manage EV data and provide personalized charging station recommendation services to EV drivers. Reference [122] proposes a blockchain-based system for assessing the battery degradation cost in EVs and generating charging station recommendations to EV drivers. The EV data (e.g., calendar aging and charging/discharging rates) is uploaded to a permissioned blockchain. The participants of the blockchain include EV users, EV charging station operators, EV aggregators, and EV mobility service providers. Among these, EV aggregators are responsible for calculating the EVs' battery degradation cost based on the EV data stored in the blockchain. The battery degradation cost is sent to the blockchain for verification and storage. Based on the degradation cost and other travel requirements, the EV mobility service providers recommend the most charging stations to a target EV driver.

The system developed in Ref. [123] perform EV charging station recommendation to EV drivers in a more privacy-preserving manner. The system is backboned by a cloud and a blockchain in which the nodes are fog computing servers. Instead of uploading the sensitive data (e.g., charging records and the EV drivers' identities and payment records) to the blockchain, only the hash of the data is uploaded. In this way, the data storage burden of the system can be significantly reduced. For ensuring data privacy, the EV drivers' sensitive data is encrypted and then sent to the cloud. For each EV driver, the cloud generates suitable fog computing servers to him/her. The EV driver selects one server from the recommendation list, and the selected fog computing server decrypts the driver's data and recommends him/her suitable charging stations. After the driver makes a choice on the charging stations, the charging data at the station will be uploaded to the fog computing server and subsequentially stored in the blockchain.

References [124, 125] focus on developing blockchain-based solutions to support V2G integration. Reference [124] develops a blockchain-based V2G integration system. The system uses a blockchain to store the data of a number of EVs that are managed by multiple EV aggregators. To launch a V2G service, the grid operator sends requests to the EV aggregators to request for energy provision. After receiving the requests, each EV aggregator retrieves the data of the EVs it the manages from the blockchain; based on the data, the EV aggregator uses certain strategies to select the most suitable EVs and make power discharging plans for the selected EVs. The EV selection result and the power discharging plans are sent to the blockchain for verification. If the information is verified as valid, the EV aggregator executes the discharging plan to provide V2G service to the grid. Reference [125] particularly focuses on the transaction authentication issues during energy trading in V2G network, where blockchain is used to verify the V2G-based energy trading transactions through asymmetric cryptography technique.

References [29, 126] describes a blockchain-based system for facilitating both V2G and vehicle-to-vehicle (V2V) integration (in the V2V scenario, an EV can share energy with other EVs through the wireless charging/discharging technology). The blockchain accepts, verifies, and stores the charging and discharging requests from the EVs and the charging stations. Based on the recorded formation in the blockchain, the EVs and the charging stations autonomously match up with each other so that the requests can be satisfied.

6 Blockchain-Based Solutions for Smart Grid Operation and Control

Besides the application scenarios in the previous sections, blockchain is also used to support many other operation and control applications in smart grids. The research efforts can be categorized as follows.

6.1 Blockchain-Based Security Constrained Economic Dispatch

Security Constrained Economic Dispatch (SCED) [127] is a fundamental task in power system operation. SCED determines the most economic power allocation (i.e., the power output scheme with the minimum generation cost) among multiple generation units subjected to a number of constraints, including the supply–demand balance constraint, power generators' operational constraints and the capacity constraint of the transmission line. SCED involves the cooperation of multiple parties, including the grid operator and power generation companies which own the generation units. In this process, there is a possibility that a party would purposely manipulate the SCED (e.g., through deliberately submitting mistaken data of the generation units) to achieve self-interests, e.g., to increase the electricity price at its region [128].

Reference [129] proposes a blockchain-based SCED scheme, which can ensure the trustworthy of the SCED result even with presence of malicious power suppliers. In the scheme, the power suppliers broadcast their power outputs to multiple nodes of a blockchain; after receiving the generation amount information, each node calculates out an adjustment plan for the suppliers' energy generation amounts and broadcasts the plan to the other nodes. Then, the nodes perform a consensus process, in which each node compares the power adjustment plan calculated by itself and the ones it receives from the other nodes. If there are malicious suppliers who deliberately sends mistaken generation amount information to the system, the agreement will not be reached among the nodes in the consensus process. As a result, the invalid generation amount information sent by malicious suppliers will not be recorded in the blockchain.

6.2 Blockchain-Based Optimal Power Flow

Optimal Power Flow (OPF) [130] is another fundamental task in power systems. OPF represents the problem of determining the most economic power generation amounts of the generation units as well as the resulted power network state (including the power flow distribution on the lines, node voltages, etc.) given a power network configuration and load information. The constraints of the power network and the generation units are also considered in OPF.

Just as the SCED process mentioned in the last section, incorrect OPF results due to misbehaviors of the different entities in the system will mislead the grid operation and may cause serious contingencies. Reference [131] develops a blockchain-based OPF scheme, which ensures the trustworthy of the OPF result through a decentralized consensus algorithm. In this scheme, the original OPF problem is decomposed into multiple sub-problems, and each one is formulated as a local OPF problem covering a certain part of the whole power network. Every node of the blockchain solves a local OPF problem and packages the result as a transaction and sends it to the blockchain. Then, a consensus process is executed. In the consensus, a leader node is selected based on a reputation algorithm, which is responsible for verifying the results of the local OPF problems. The leader node packages the valid results to a block and broadcasts it to the system. The other nodes then further verify the results that has been verified by the leader node. Once the results have been confirmed to be valid by majority of the nodes, they will be recorded into the blockchain.

6.3 Blockchain-Based Grid Topology Change Identification

With the high complexity that is resulted by the factors such as widespread deployment of distributed energy resources and increasing energy demand, unexpected topology changes would happen in smart grids. Promptly identifying the grid's topological anomalies can provide support to making countermeasures and maintaining the grid's secure and reliable operation. Reference [132] develops a blockchain-based solution for identifying power distribution networks' topology changes from the data collected by PMUs. In the system, a distribution network is divided into multiple areas, and PMUs are deployed in each area to monitor the bus voltages. Each node of the blockchain acts as a manager of a specific area, which collects and analyzes the data from the PMUs in that area. If any topological anomalies are identified, the node will send the information about the changed topology into smart contracts, which will be triggered to verify the changed topology information by comparing it with the historical topology information that has been stored in the blockchain. If the new topology information passes the verification, it will be stored into the blockchain, which represents the distribution network's topology has been updated.

6.4 Blockchain-Based Microgrid Energy Management

With the wider acceptance and penetration of distributed renewable energy sources, the concept of microgrid has gained extensive popularity [133–135]. The use of the term 'micro' is to distinguish a microgrid from the main grid. A microgrid aggregates and manages a variety of energy resources in a certain geographical area through information and physical facilities. These energy resources include both energy generation sources and energy loads. The connection between a microgrid and the main grid occurs at common coupling points. Conventionally, energy management of microgrids is performed by central Energy Management Systems (EMSs). Such a centralized energy management pattern is associated with the challenges of high maintenance cost for the central EMS, single-point-of-failure risk, and inefficiency in managing the large volume of data from the distributed energy resources. In the literature [27, 97, 136, 137], blockchain-based distributed energy management solutions have been developed for microgrids.

6.4.1 Secure Operation of Microgrids

References [136, 137] focus on utilizing blockchain to facilitate secure operation for microgrids. Reference [136] develops a blockchain-based distributed frequency control scheme for microgrids. In the scheme, four distributed control systems (DCSs) are set up for collaboratively managing a microgrid's frequency. The DCSs also act as the nodes of a blockchain. Each DCS firstly measures the frequency locally; the measured frequency values are set to the blockchain to trigger a smart contract that is responsible for comparing the locally measured frequency values and make frequency regulation decisions. The smart contract then generates a transaction to the blockchain system, which contains the information of the frequency deviation between adjacent areas in the microgrid and the frequency regulation instructions. The transaction is stored in the blockchain, and the DCSs can retrieve the frequency regulation instructions to perform frequency regulation actions.

Reference [137] designs a blockchain-based scheme to ensure the physical constraints of the microgrid network (e.g., the constraints of node voltage magnitude and transmission line capacity) are not violated in the microgrid's operation. In the scheme, the energy sources in the microgrid send the amounts of power they plan to generate to the blockchain. Then, the microgrid operator retrieves the information and checks if there are any physical constraints violated. If no constraints violated, the microgrid operator will send an agreement signal to a trigger a smart contract, and the latter will send broadcast a confirmation message to the energy sources to allow them to arrange electricity generation; otherwise, the microgrid operator will make power output adjustment instructions and trigger the smart contract to broadcast the instructions to the energy sources.

6.4.2 Cooperation of Multiple Microgrids

Since a microgrid typically aggregates the energy resources in a specific area, it is possible to enable multiple microgrids to communicate and exchange energy with each other. References [27, 97] study using blockchain to support the cooperation of multiple microgrids. Reference [97] proposes a blockchain-based decentralized system for dispatching power outputs among multiple microgrids. For each microgrid, the users in the microgrid uploads their power production and consumption data to the microgrid EMS; the EMS of each microgrid sends the total power production and consumption data of the microgrid to a blockchain. A smart contract then will be triggered to solve a power dispatch problem and determines the power output each microgrid should provide. The power dispatch result is then stored in the blockchain. The blockchain is managed by a limited number of nodes that are configured with strong computing power, making the smart contract be capable of performing compute-intensive power dispatch computation.

In the multi-microgrid cooperation scenario, energy exchange happens in two levels: among the internal energy entities of a microgrid and among the microgrids. Reference [27] designs a hierarchical blockchain structure for facilitating the trustable energy exchange in both levels. In the lower level, a blockchain is set up for each microgrid, which is collaboratively managed by the microgrid EMS and the prosumers in the microgrid. Such a local blockchain is responsible for recording the internal energy exchange transactions among the prosumers in the microgrid. In the upper level, another blockchain is set up, which is managed by the microgrid EMSs. This blockchain is responsible for recording the energy exchange transactions among the microgrids.

6.5 *Blockchain-Based Virtual Power Plant Energy Management*

Virtual Power Plant (VPP) [138, 139] is another paradigm of aggregating distributed energy resources. A VPP refers to an energy entity that aggregates geographically dispersed energy resources and operates like a conventional power plant. Compared with a microgrid where the energy resources are usually located in a same area, the energy resources in a VPP can span wide geographical areas and they can be managed by a centralized EMS.

Blockchain has been utilized to support the management of the dispersed energy resources in VPPs. In Ref. [140] develops a smart contract-based mechanism to enable the VPP operator to directly control the energy resources. The owners of the energy resources sign agreements with the VPP operator on energy resource control, and the agreements are stored in a smart contract. When the VPP operator operates the energy resources, the operation data is stored in the blockchain. As per the operation data, the smart contract is triggered to automatically transfer the compensation payment to the energy resource owners. In the system in Ref. [141], the users in the VPP firstly share their energy production and consumption schedules to a blockchain system; then, multiple smart contracts execute some algorithms to: (i) determine energy trading pairs among the users, and (ii) dispatch the energy storage units to store the surplus energy produced from the internal renewable energy sources and/or from the grid.

In particular, Refs. [142, 143] develop two blockchain-based interaction applications between VPPs and the grid. Reference [142] proposes a dynamic tax rate scheme for VPPs. The dynamic tax rate means when a user in a VPP sells energy to another, the seller needs to pay a certain tax to the grid, and the tax rate depends on the predicted power supply and demand of the grid. In the system, a blockchain is used to store the power generation and consumption data of the users; based on the data, each node generates the power supply and demand data in a certain area and determines the tax rate. The tax payments of the users are verified by the blockchain's consensus process and are stored in the blockchain. Reference [143] uses blockchain to facilitate VPPs to provide load shifting service to the grid. At first, the grid sends the expected power load information to a blockchain system; based on the information, the VPP operator controls the energy resources in the VPP to satisfy the instruction, and the actual load of the VPP is verified and stored in the blockchain as well.

For the systems in the above literature [140–143], the VPP operator (or VPP EMS) plays an important role, as it takes the most work of making energy management decisions. References [144, 145] designs schemes to make VPPs operate in a more decentralized manner. In the scheme, a VPP is divided into multiple sub-VPPs. Each sub-VPP is managed by an EMS; the EMS manages the energy resources in a specific sub-area of the VPP and acts as a node of a blockchain. The EMSs of the sub-VPPs upload the operation data of the energy resources to a blockchain. The signals from the grid (e.g., load shifting requests) will be sent to the blockchain, which will trigger smart contracts to make control plans to the energy resources subjected to the grid-side signals.

7 Conclusion

Due to its decentralization feature and its capability of ensuring data immutability and traceability, the blockchain technology has shown its large application potential in energy systems. In this chapter, we review the state-of-the-art of the application of blockchain in different areas in smart grids in recent years (between 2020 to May 2022). This chapter is devoted to help the reader have basic understanding on how blockchain can support the operation of modern smart grids and is expected to provide a reference to students, researchers and engineers in the related areas.

The trust the blockchain technology can provide to energy systems (and other domains) largely comes from its PoW-based consensus mechanism; through setting computationally difficult puzzles, a blockchain can eliminate the intension of a rationale attacker to maliciously generate blocks due to the high cost. However, such a mechanism is also associated with large electricity consumption used in generating the proof of work (e.g., solving mathematically difficult puzzles)—this is essentially conflict with the energy efficient principle of smart grids. As reviewed in this chapter, research efforts have been conducted to design less energy-hungry consensus mechanisms for blockchain, but the cost of the reduced energy consumption of consensus is the reduced decentralization degree and/or the less cyber security of the system. In future, how to balance the energy efficiency, cyber security and trust, and throughput of a blockchain system in the smart grid environment is still a problem that is worthy of investigation. In addition, smart grids are still under rapid development and many new applications and technologies are being developed in this paradigm. In this context, it is necessary to put more efforts in exploring new blockchain-supported smart grid applications and fully underpinning the ability of blockchain for future energy systems.

Acknowledgements This work is supported by the Australian Research Council through a Discovery Project (DP220103881).

References

1. Beagley J et al (2021) Assessing the inclusion of health in national climate commitments: towards accountability for planetary health. J Clim Change Health 5:100085–100085
2. Tuballa ML, Abundo ML (2016) A review of the development of smart grid technologies. Renew Sustain Energy Rev 59:710–725
3. Rehmani MH, Reisslein M, Rachedi A, Erol-Kantarci M, Radenkovic M (2018) Integrating renewable energy resources into the smart grid: recent developments in information and communication technologies. IEEE Trans Ind Inform 14(7):2814–2825
4. Aceto G, Persico V, Pescapé A (2019) A survey on information and communication technologies for industry 4.0: state-of-the-art, taxonomies, perspectives, and challenges. IEEE Commun Surv Tutor 21(4):3467–3501
5. Aziz A, Khalid S, Mustafa M, Shareef H, Aliyu G (2013) Artificial intelligent meter development based on advanced metering infrastructure technology. Renew Sustain Energy Rev 27:191–197
6. Belotti M, Božić N, Pujolle G, Secci S (2019) A vademecum on blockchain technologies: when, which, and how. IEEE Commun Surv Tutor 21(4):3796–3838
7. Zhuang P, Zamir T, Liang H (2020) Blockchain for cybersecurity in smart grid: a comprehensive survey. IEEE Trans Ind Inform 17(1):3–19
8. Kirli D et al (2022) Smart contracts in energy systems: a systematic review of fundamental approaches and implementations. Renew Sustain Energy Rev 158:112013
9. Mollah MB et al (2020) Blockchain for future smart grid: a comprehensive survey. IEEE Internet Things J 8(1):18–43
10. Di Silvestre ML, Gallo P, Ippolito MG, Sanseverino ER, Zizzo G (2018) A technical approach to the energy blockchain in microgrids. IEEE Trans Ind Inform 14(11):4792–4803
11. Wang N et al (2019) When energy trading meets blockchain in electrical power system: the state of the art. Appl Sci 9(8):1561
12. Woo J, Fatima R, Kibert CJ, Newman RE, Tian Y, Srinivasan RS (2021) Applying blockchain technology for building energy performance measurement, reporting, and verification (MRV) and the carbon credit market: a review of the literature. Build Environ 205:108199
13. Júnior CAR, Sanseverino ER, Gallo P, Koch D, Schweiger H-G, Zanin H (2022) Blockchain review for battery supply chain monitoring and battery trading. Renew Sustain Energy Rev 157:112078
14. Andoni M et al (2019) Blockchain technology in the energy sector: a systematic review of challenges and opportunities. Renew Sustain Energy Rev 100:143–174
15. Nakamoto S (2008) Bitcoin: a peer-to-peer electronic cash system. Decentralized Business Review, p 21260

16. Gupta SS (2017) Blockchain. IBM online: http://www.ibm.com
17. Rasjid ZE, Soewito B, Witjaksono G, Abdurachman E (2017) A review of collisions in cryptographic hash function used in digital forensic tools. Procedia Comput Sci 116:381–392
18. Gervais A, Karame GO, Wüst K, Glykantzis V, Ritzdorf H, Capkun S (2016) On the security and performance of proof of work blockchains. In: Proceedings of the 2016 ACM SIGSAC conference on computer and communications security, pp 3–16
19. King S, Nadal S (2012) Ppcoin: peer-to-peer crypto-currency with proof-of-stake. Self-published paper, vol 19, no 1, Aug 2012
20. Castro M, Liskov B (1999) Practical Byzantine fault tolerance. OsDI 99(1999):173–186
21. Buterin V (2014) A next-generation smart contract and decentralized application platform. White Pap 3(37):2.1
22. Morello R, Mukhopadhyay SC, Liu Z, Slomovitz D, Samantaray SR (2017) Advances on sensing technologies for smart cities and power grids: a review. IEEE Sens J 17(23):7596–7610
23. Asghar MR, Dán G, Miorandi D, Chlamtac I (2017) Smart meter data privacy: a survey. IEEE Commun Surv Tutor 19(4):2820–2835
24. Karthikeyan SP, Raglend IJ, Kothari DP (2013) A review on market power in deregulated electricity market. Int J Electr Power Energy Syst 48:139–147
25. Hongling L, Chuanwen J, Yan Z (2008) A review on risk-constrained hydropower scheduling in deregulated power market. Renew Sustain Energy Rev 12(5):1465–1475
26. Morstyn T, Farrell N, Darby SJ, McCulloch MD (2018) Using peer-to-peer energy-trading platforms to incentivize prosumers to form federated power plants. Nat Energy 3(2):94–101. https://doi.org/10.1038/s41560-017-0075-y
27. Huang X, Zhang Y, Li D, Han L (2022) A solution for bi-layer energy trading management in microgrids using multi-blockchain. IEEE Internet Things J 1
28. Jogunola O et al (2020) Consensus algorithms and deep reinforcement learning in energy market: a review. IEEE Internet Things J 8(6):4211–4227
29. Abishu HN, Seid AM, Yacob YH, Ayall T, Sun G, Liu G (2022) Consensus mechanism for blockchain-enabled vehicle-to-vehicle energy trading in the internet of electric vehicles. IEEE Trans Veh Technol 71(1):946–960
30. AlAshery MK et al (2020) A blockchain-enabled multi-settlement quasi-ideal peer-to-peer trading framework. IEEE Trans Smart Grid 12(1):885–896
31. Chen S et al (2021) A trusted energy trading framework by marrying blockchain and optimization. Adv Appl Energy 2:100029
32. Hamouda MR, Nassar ME, Salama M (2020) A novel energy trading framework using adapted blockchain technology. IEEE Trans Smart Grid 12(3):2165–2175
33. Esmat A, de Vos M, Ghiassi-Farrokhfal Y, Palensky P, Epema D (2021) A novel decentralized platform for peer-to-peer energy trading market with blockchain technology. Appl Energy 282:116123
34. Wu Y, Zhang X, Sun H (2021) A multi-time-scale autonomous energy trading framework within distribution networks based on blockchain. Appl Energy 287:116560
35. Vieira G, Zhang J (2021) Peer-to-peer energy trading in a microgrid leveraged by smart contracts. Renew Sustain Energy Rev 143:110900
36. Yang J, Paudel A, Gooi HB, Nguyen HD (2021) A proof-of-stake public blockchain based pricing scheme for peer-to-peer energy trading. Appl Energy 298:117154
37. Yang J, Dai J, Gooi HB, Nguyen HD, Wang P (2022) Hierarchical blockchain design for distributed control and energy trading within microgrids. IEEE Trans Smart Grid
38. Yang Q, Wang H (2020) Blockchain-empowered socially optimal transactive energy system: framework and implementation. IEEE Trans Ind Inform 17(5):3122–3132
39. Zhang C, Yang T, Wang Y (2021) Peer-to-peer energy trading in a microgrid based on iterative double auction and blockchain. Sustain Energy Grids Netw 27:100524
40. AlSkaif T, Crespo-Vazquez JL, Sekuloski M, van Leeuwen G, Catalão JP (2021) Blockchain-based fully peer-to-peer energy trading strategies for residential energy systems. IEEE Trans Ind Inform 18(1):231–241

41. Abdella J, Tari Z, Anwar A, Mahmood A, Han F (2021) An architecture and performance evaluation of blockchain-based peer-to-peer energy trading. IEEE Trans Smart Grid 12(4):3364–3378
42. Jamil F, Iqbal N, Ahmad S, Kim D (2021) Peer-to-peer energy trading mechanism based on blockchain and machine learning for sustainable electrical power supply in smart grid. IEEE Access 9:39193–39217
43. Pradhan NR, Singh AP, Kumar N, Hassan M, Roy D (2021) A flexible permission ascription (FPA) based blockchain framework for peer-to-peer energy trading with performance evaluation. IEEE Trans Ind Inform
44. Muzumdar A, Modi C, Madhu G, Vyjayanthi C (2021) A trustworthy and incentivized smart grid energy trading framework using distributed ledger and smart contracts. J Netw Comput Appl 183:103074
45. Yahaya AS, Javaid N, Almogren A, Ahmed A, Gulfam SM, Radwan A (2021) A two-stage privacy preservation and secure peer-to-peer energy trading model using blockchain and cloud-based aggregator. IEEE Access 9:143121–143137
46. Guo J, Ding X, Wu W (2020) A blockchain-enabled ecosystem for distributed electricity trading in smart city. IEEE Internet Things J 8(3):2040–2050
47. Wang B et al (2021) Design of a privacy-preserving decentralized energy trading scheme in blockchain network environment. Int J Electr Power Energy Syst 125:106465
48. Nykyri M, Kärkkäinen TJ, Levikari S, Honkapuro S, Annala S, Silventoinen P (2022) Blockchain-based balance settlement ledger for energy communities in open electricity markets. Energy 124180
49. Wang H, Wu Z, Li Y, Yan Z, Ma J (2021) Architecture design and application of distributed power trading system based on blockchain asynchronous consensus. In: 2021 4th international conference on advanced electronic materials, computers and software engineering (AEMCSE). IEEE, pp 35–41
50. Tan W et al (2022) Blockchain-based distributed power transaction mechanism considering credit management. Energy Rep 8:565–572
51. Wongthongtham P, Marrable D, Abu-Salih B, Liu X, Morrison G (2021) Blockchain-enabled peer-to-peer energy trading. Comput Electr Eng 94:107299
52. Khorasany M, Dorri A, Razzaghi R, Jurdak R (2021) Lightweight blockchain framework for location-aware peer-to-peer energy trading. Int J Electr Power Energy Syst 127:106610
53. Dorri A, Luo F, Karumba S, Kanhere S, Jurdak R, Dong ZY (2021) Temporary immutability: a removable blockchain solution for prosumer-side energy trading. J Netw Comput Appl 180:103018
54. Oprea S-V, Bâra A, Diaconita V (2022) A motivational local trading framework with 2-round auctioning and settlement rules embedded in smart contracts for a small citizen energy community. Renew Energy
55. Wang L, Ma Y, Zhu L, Wang X, Cong H, Shi T (2022) Design of integrated energy market cloud service platform based on blockchain smart contract. Int J Electr Power Energy Syst 135:107515
56. Han D, Zhang C, Ping J, Yan Z (2020) Smart contract architecture for decentralized energy trading and management based on blockchains. Energy 199:117417
57. Wang T, Guo J, Ai S, Cao J (2021) RBT: a distributed reputation system for blockchain-based peer-to-peer energy trading with fairness consideration. Appl Energy 295:117056
58. Zhou K, Chong J, Lu X, Yang S (2021) Credit-based peer-to-peer electricity trading in energy blockchain environment. IEEE Trans Smart Grid 13(1):678–687
59. Li M, Hu D, Lal C, Conti M, Zhang Z (2020) Blockchain-enabled secure energy trading with verifiable fairness in industrial internet of things. IEEE Trans Ind Inform 16(10):6564–6574
60. Gao G et al (2021) FogChain: a blockchain-based peer-to-peer solar power trading system powered by fog AI. IEEE Internet Things J
61. Yahaya AS et al (2022) Blockchain based secure energy trading with mutual verifiable fairness in a smart community. IEEE Trans Ind Inform 1

62. Al Sadawi A, Madani B, Saboor S, Ndiaye M, Abu-Lebdeh G (2021) A comprehensive hierarchical blockchain system for carbon emission trading utilizing blockchain of things and smart contract. Technol Forecast Soc Change 173:121124
63. Effah D, Chunguang B, Appiah F, Agbley BLY, Quayson M (2021) Carbon emission monitoring and credit trading: the blockchain and IOT approach. In: 2021 18th international computer conference on wavelet active media technology and information processing (ICCWAMTIP). IEEE, pp 106–109
64. Richardson A, Xu J (2020) Carbon trading with blockchain. In: Mathematical research for blockchain economy. Springer, pp 105–124
65. Kim S-K, Huh J-H (2020) Blockchain of carbon trading for UN sustainable development goals. Sustainability 12(10):4021
66. Muzumdar A, Modi C, Vyjayanthi C (2022) A permissioned blockchain enabled trustworthy and incentivized emission trading system. J Clean Prod 349:131274
67. Patel D, Britto B, Sharma S, Gaikwad K, Dusing Y, Gupta M (2020) Carbon credits on blockchain. In: 2020 international conference on innovative trends in information technology (ICITIIT). IEEE, pp 1–5
68. Muzumdar CM, Vyjayanthi C (2022) A permissioned blockchain enabled trustworthy and incentivized emission trading system. J Clean Prod 349:131274
69. Zhao N, Sheng Z, Yan H (2021) Emission trading innovation mechanism based on blockchain. Chin J Popul Resour Environ 19(4):369–376
70. Lu Y et al (2022) STRICTs: a blockchain-enabled smart emission cap restrictive and carbon permit trading system. Appl Energy 313:118787
71. Lakshmi G, Thiyagarajan G (2021) Decentralized energy to power rural homes through smart contracts and carbon credit. In: 2021 7th international conference on electrical energy systems (ICEES). IEEE, pp 280–283
72. Lyons RK, Viswanath-Natraj G (2020) What keeps stablecoins stable? National Bureau of Economic Research
73. Liu Y, Tsyvinski A (2021) Risks and returns of cryptocurrency. Rev Financ Stud 34(6):2689–2727
74. Toderean L et al (2021) A lockable ERC20 token for peer to peer energy trading. arXiv preprint arXiv:2111.04467
75. Ding S, Zeng J, Hu Z, Yang Y (2021) A peer-2-peer management and secure policy of the energy internet in smart microgrids. IEEE Trans Ind Inform
76. Energy Web (2020) EW-DOS: the energy web decentralized operating system. In: An open-source technology stack to accelerate the energy transition. Part II, technology detail
77. Waters L, Tal I (2021) CERCoin: carbon tracking enabling blockchain system for electric vehicles. In: 2021 IEEE 21st international conference on software quality, reliability and security companion (QRS-C). IEEE, pp 622–629
78. Eckert J, Lopez D, Azevedo CL, Farooq B (2020) A blockchain-based user-centric emission monitoring and trading system for multi-modal mobility. IEEE, pp 328–334
79. Mehdinejad M, Shayanfar H, Mohammadi-Ivatloo B (2022) Decentralized blockchain-based peer-to-peer energy-backed token trading for active prosumers. Energy 244:122713
80. Truby J, Brown RD, Dahdal A, Ibrahim I (2022) Blockchain, climate damage, and death: policy interventions to reduce the carbon emissions, mortality, and net-zero implications of non-fungible tokens and Bitcoin. Energy Res Soc Sci 88:102499
81. Munoz MF, Zhang K, Amara F (2022) ZipZap: a blockchain solution for local energy trading. arXiv preprint arXiv:2202.13450
82. Karandikar N, Chakravorty A, Rong C (2021) Blockchain based transaction system with fungible and non-fungible tokens for a community-based energy infrastructure. Sensors 21(11):3822
83. Wang Q, Li R, Wang Q, Chen S (2021) Non-fungible token (NFT): overview, evaluation, opportunities and challenges. arXiv preprint arXiv:2105.07447
84. Gadde PH, Biswal M, Brahma S, Huiping C (2016) Efficient compression of PMU data in WAMS. IEEE Trans Smart Grid 7(5):2406–2413. https://doi.org/10.1109/TSG.2016.2536718

85. Arenas-Martínez M et al (2010) A comparative study of data storage and processing architectures for the smart grid. In: 2010 first ieee international conference on smart grid communications. IEEE, pp 285–290
86. Zhang S, Rong J, Wang B (2020) A privacy protection scheme of smart meter for decentralized smart home environment based on consortium blockchain. Int J Electr Power Energy Syst 121:106140
87. Olivares-Rojas JC, Reyes-Archundia E, Gutiérrez-Gnecchi JA, Molina-Moreno I, Cerda-Jacobo J, Méndez-Patiño A (2021) A transactive energy model for smart metering systems using blockchain. CSEE J Power Energy Syst 7(5):943–953
88. Lu W, Ren Z, Xu J, Chen S (2021) Edge blockchain assisted lightweight privacy-preserving data aggregation for smart grid. IEEE Trans Netw Serv Manage 18(2):1246–1259
89. Sedlmeir J, Völter F, Strüker J (2021) The next stage of green electricity labeling: using zero-knowledge proofs for blockchain-based certificates of origin and use. ACM SIGEnergy Energy Inform Rev 1(1):20–31
90. Zhang X, Song Z, Moshayedi AJ (2022) Security scheduling and transaction mechanism of virtual power plants based on dual blockchains. J Cloud Comput 11(1):1–26
91. Kumari A, Tanwar S (2021) A reinforcement learning-based secure demand response scheme for smart grid system. IEEE Internet Things J
92. Kumari A, Tanwar S (2020) A data analytics scheme for security-aware demand response management in smart grid system. In: 2020 IEEE 7th Uttar Pradesh section international conference on electrical, electronics and computer engineering (UPCON). IEEE, pp 1–6
93. Zheng X, Lu J, Sun S, Kiritsis D (2020) Decentralized industrial IoT data management based on blockchain and IPFS. In: IFIP international conference on advances in production management systems. Springer, pp 222–229
94. Kumari A, Trivedi M, Tanwar S, Sharma G, Sharma R (2022) SV2G-ET: a secure vehicle-to-grid energy trading scheme using deep reinforcement learning. Int Trans Electr Energy Syst 2022
95. Daniel E, Tschorsch F (2022) IPFS and friends: a qualitative comparison of next generation peer-to-peer data networks. IEEE Commun Surv Tutor
96. Benet J (2014) IPFS-content addressed, versioned, P2P file system. arXiv preprint arXiv: 1407.3561
97. Luo X, Xue K, Xu J, Sun Q, Zhang Y (2021) Blockchain based secure data aggregation and distributed power dispatching for microgrids. IEEE Trans Smart Grid 12(6):5268–5279
98. Fontaine C, Galand F (2007) A survey of homomorphic encryption for nonspecialists. EURASIP J Inf Secur 2007:1–10
99. Guan Z, Zhou X, Liu P, Wu L, Yang W (2021) A blockchain based dual side privacy preserving multi party computation scheme for edge enabled smart grid. IEEE Internet Things J
100. Du W, Atallah MJ (2001) Secure multi-party computation problems and their applications: a review and open problems. In: Proceedings of the 2001 workshop on new security paradigms, pp 13–22
101. Miyamae T et al (2021) ZGridBC: zero-knowledge proof based scalable and private blockchain platform for smart grid. In: 2021 IEEE international conference on blockchain and cryptocurrency (ICBC). IEEE, pp 1–3
102. Wang W, Huang H, Zhang L, Su C (2021) Secure and efficient mutual authentication protocol for smart grid under blockchain. Peer-to-Peer Netw Appl 14(5):2681–2693
103. Zhong Y et al (2021) Distributed blockchain-based authentication and authorization protocol for smart grid. Wireless Commun Mob Comput 2021
104. Chandra S, Paira S, Alam SS, Sanyal G (2014) A comparative survey of symmetric and asymmetric key cryptography. In: 2014 international conference on electronics, communication and computational engineering (ICECCE). IEEE, pp 83–93
105. Goldreich O, Micali S, Wigderson A (1991) Proofs that yield nothing but their validity or all languages in NP have zero-knowledge proof systems. J ACM 38(3):690–728
106. Rackoff C, Simon DR (1991) Non-interactive zero-knowledge proof of knowledge and chosen ciphertext attack. In: Annual international cryptology conference. Springer, pp 433–444

107. Goldreich O, Oren Y (1994) Definitions and properties of zero-knowledge proof systems. J Cryptol 7(1):1–32
108. Samadi M, Schriemer H, Ruj S, Erol-Kantarci M (2021) Energy blockchain for demand response and distributed energy resource management. In: 2021 IEEE international conference on communications, control, and computing technologies for smart grids (SmartGridComm). IEEE, pp 425–431
109. Ma R, Yi Z, Xiang Y, Shi D, Xu C, Wu H (2022) A blockchain-enabled demand management and control framework driven by deep reinforcement learning. IEEE Trans Ind Electron
110. Lucas A, Geneiatakis D, Soupionis Y, Nai-Fovino I, Kotsakis E (2021) Blockchain technology applied to energy demand response service tracking and data sharing. Energies 14(7):1881
111. Tsao Y-C, Thanh V-V, Wu Q (2021) Sustainable microgrid design considering blockchain technology for real-time price-based demand response programs. Int J Electr Power Energy Syst 125:106418
112. Alam MR, Reaz MBI, Ali MAM (2012) A review of smart homes—past, present, and future. IEEE Trans Syst Man Cybern Part C (Appl Rev) 42(6):1190–1203
113. Chan M, Estève D, Escriba C, Campo E (2008) A review of smart homes—present state and future challenges. Comput Methods Programs Biomed 91(1):55–81
114. Ren Y et al (2021) Multiple cloud storage mechanism based on blockchain in smart homes. Future Gener Comput Syst 115:304–313
115. Yang Q, Wang H (2021) Privacy-preserving transactive energy management for IoT-aided smart homes via blockchain. IEEE Internet Things J 8(14):11463–11475
116. Pazhev G, Spasov G, Shopov M, Petrova G (2020) On the use of blockchain technologies in smart home applications. IOP Conf Ser Mater Sci Eng 878(1):012023. IOP Publishing
117. Chen C, Wang J, Kishore S (2014) A distributed direct load control approach for large-scale residential demand response. IEEE Trans Power Syst 29(5):2219–2228
118. Lee Y, Rathore S, Park JH, Park JH (2020) A blockchain-based smart home gateway architecture for preventing data forgery. HCIS 10(1):1–14
119. Kempton W, Tomić J (2005) Vehicle-to-grid power implementation: from stabilizing the grid to supporting large-scale renewable energy. J Power Sources 144(1):280–294
120. Liu C, Chau K, Wu D, Gao S (2013) Opportunities and challenges of vehicle-to-home, vehicle-to-vehicle, and vehicle-to-grid technologies. Proc IEEE 101(11):2409–2427
121. Florea BC, Taralunga DD (2020) Blockchain IoT for smart electric vehicles battery management. Sustainability 12(10):3984
122. Gowda SN, Eraqi BA, Nazaripouya H, Gadh R (2021) Assessment and tracking electric vehicle battery degradation cost using blockchain. In: 2021 IEEE power & energy society innovative smart grid technologies conference (ISGT). IEEE, pp 1–5
123. Li H, Han D, Tang M (2020) A privacy-preserving charging scheme for electric vehicles using blockchain and fog computing. IEEE Syst J 15(3):3189–3200
124. Aggarwal S, Kumar N (2021) A consortium blockchain-based energy trading for demand response management in vehicle-to-grid. IEEE Trans Veh Technol 70(9):9480–9494
125. Aggarwal S, Kumar N, Gope P (2020) An efficient blockchain-based authentication scheme for energy-trading in V2G networks. IEEE Trans Ind Inform 17(10):6971–6980
126. Luo L, Feng J, Yu H, Sun G (2021) Blockchain-enabled two-way auction mechanism for electricity trading in internet of electric vehicles. IEEE Internet Things J
127. Vargas LS, Quintana VH, Vannelli A (1993) A tutorial description of an interior point method and its applications to security-constrained economic dispatch. IEEE Trans Power Syst 8(3):1315–1324
128. Zhao C, He J, Cheng P, Chen J (2017) Analysis of consensus-based distributed economic dispatch under stealthy attacks. IEEE Trans Ind Electron 64(6):5107–5117
129. Chen S, Zhang L, Yan Z, Shen Z (2021) A distributed and robust security-constrained economic dispatch algorithm based on blockchain. IEEE Trans Power Syst 37(1):691–700
130. Huneault M, Galiana FD (1991) A survey of the optimal power flow literature. IEEE Trans Power Syst 6(2):762–770

131. Foti M, Mavromatis C, Vavalis M (2021) Decentralized blockchain-based consensus for optimal power flow solutions. Appl Energy 283:116100
132. Li X, Han B, Li G, Luo L, Wang K, Jiang X (2021) Dynamic topology awareness in active distribution networks using blockchain-based state estimations. IEEE Trans Power Syst 36(6):5185–5197
133. Olivares DE et al (2014) Trends in microgrid control. IEEE Trans Smart Grid 5(4):1905–1919
134. Lasseter RH, Paigi P (2004) Microgrid: a conceptual solution. In: 2004 IEEE 35th annual power electronics specialists conference (IEEE Cat. No. 04CH37551), vol 6. IEEE, pp 4285–4290
135. Mengelkamp E, Gärttner J, Rock K, Kessler S, Orsini L, Weinhardt C (2018) Designing microgrid energy markets: a case study: the Brooklyn microgrid. Appl Energy 210:870–880
136. Yang J, Dai J, Gooi HB, Nguyen H, Paudel A (2022) A proof-of-authority blockchain based distributed control system for islanded microgrids. IEEE Trans Ind Inform
137. Wang B, Liu H, Zhang S (2022) A privacy protection scheme for electricity transactions in the microgrid day-ahead market based on consortium blockchain. Int J Electr Power Energy Syst 141:108144
138. Pudjianto D, Ramsay C, Strbac G (2007) Virtual power plant and system integration of distributed energy resources. IET Renew Power Gener 1(1):10–16
139. Bhuiyan EA, Hossain MZ, Muyeen S, Fahim SR, Sarker SK, Das SK (2021) Towards next generation virtual power plant: technology review and frameworks. Renew Sustain Energy Rev 150:111358
140. Yang Q, Wang H, Wang T, Zhang S, Wu X, Wang H (2021) Blockchain-based decentralized energy management platform for residential distributed energy resources in a virtual power plant. Appl Energy 294:117026
141. Seven S, Yao G, Soran A, Onen A, Muyeen S (2020) Peer-to-peer energy trading in virtual power plant based on blockchain smart contracts. IEEE Access 8:175713–175726
142. Hassan MU, Rehmani MH, Chen J (2021) VPT: privacy preserving energy trading and block mining mechanism for blockchain based virtual power plants. arXiv preprint arXiv:2102.01480
143. Mnatsakanyan A, Albeshr H, Al Marzooqi A, Bilbao E (2020) Blockchain-integrated virtual power plant demonstration. In: 2020 2nd international conference on smart power & internet energy systems (SPIES). IEEE, pp 172–175
144. Mathew R, Mehbodniya A, Ambalgi AP, Murali M, Sahay KB, Babu DV (2022) In a virtual power plant, a blockchain-based decentralized power management solution for home distributed generation. Sustain Energy Technol Assess 49:101731
145. Cioara T, Antal M, Mihailescu VT, Antal CD, Anghel IM, Mitrea D (2021) Blockchain-based decentralized virtual power plants of small prosumers. IEEE Access 9:29490–29504

Client Selection Frameworks Within Federated Machine Learning: The Current Paradigm

Lincoln Best, Ernest Foo, Hui Tian, and Zahra Jadidi

Abstract Organisations are increasingly looking for ways to further utilise big data and the benefits that come with this. Previously, this role has been taken by traditional machine learning algorithms. However, these have drawbacks such as computation cost and privacy issues. Federated machine learning (FML) seeks to remedy the downfalls of traditional machine learning. Client selection is one way in which to further improve FML, as which clients that are chosen, and how they operate are a core part of its operation. This paper proposes a potential better way to operate a client selection framework, after reviewing the current literature within academia.

Keywords Federated machine learning · Client selection framework · Cyber security

1 Introduction

Companies are increasingly looking at ways to utilise the large amount of data generated by users. This data is often used to improve products and services, as well as the welfare of customers. This increase in data generation has subsequently made the associated devices, more of target for malicious actors [6]. This means that there is now a greater need for harnessing data in a safe and secure manner. This is where federated machine learning (FML) comes in. FML seeks to ensure that any computation that takes place on the data, within the clients are done so with privacy in mind.

L. Best (✉) · E. Foo · H. Tian · Z. Jadidi
School of Information and Communication Technology, Griffith University, Brisbane, Australia
e-mail: lincoln.best@griffith.edu.au

E. Foo
e-mail: e.foo@griffith.edu.au

H. Tian
e-mail: hui.tian@griffith.edu.au

Z. Jadidi
e-mail: z.jadidi@griffith.edu.au

S. Pal et al. (eds.), *Emerging Smart Technologies for Critical Infrastructure*,
Smart Sensors, Measurement and Instrumentation 44,
https://doi.org/10.1007/978-3-031-29845-5_3

FML operates by only allowing the model to be passed between different nodes, whilst the data that is being computed remains within the client. This ensures that the risk of the identifiable data being intercepted in reduced substantially [16].

2 Literature Review

This literature review will first introduce the concept of FML by delving into the composition of the algorithm. This literature review will also look at the modules and the layers that comprise a standard FML algorithm. Next, we look at the different types of FML (horizontal, vertical and transfer learning). After this, we discuss and analyse multiple different client selection frameworks (CSF's) within academia, before going on to look at applications for FML and that relation to a strong CSF. Lastly, we look at the future applications and the current gaps within academia. This review will begin by providing an overview of FML as a concept (Fig. 1).

2.1 Federated Machine Learning Overview

FML is a technological framework that allows a machine learning model to be created and applied to the generated data in a collective fashion that is distributed across

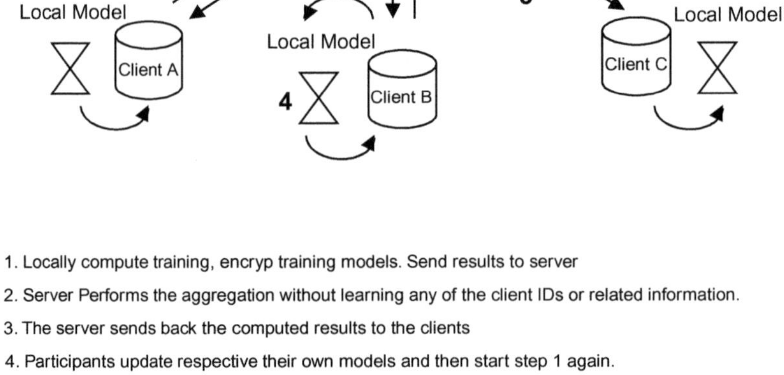

Fig. 1 An diagram of the general operation of federated machine learning

different organisations and devices [16]. Whilst this framework operates, it actively aims to preserve privacy, security and meet particular security needs regarding data access and use. FML has some distinct advantages over its more traditional counterparts. These include but are not limited to scalability, low throughput improving accuracy, improving training time and costs, ensuring privacy and security, and minimising data usage. Scalability for FML refers to the notion of each client device being able to learn from one another, thus making the entire network able to scale in size. The low throughput idea means that FML as a concept creates local models that aide in reduce latency which therefore results in lower power consumption compared to the traditional centralised, single model machine learning algorithms. The accuracy improvement aspect of FML comes from training local models, which provide a more customised approach to data, all whilst operating simultaneously. The reduction in time and training cost happens as the local models are smaller and more agile then the standard centralised models as previously mentioned. The Privacy and security improvements utilised by FML is largely due to the fact that the data never leaves the client or database it is stored in, the model is transmitted instead, this works well with the General Data Protection Regulation (GDPR) which is currently utilised by the European Union. Finally, data minimisation utilises the data minimisation tenet in order to ensure that only the learned model is processed by a central server. Compared to the raw data which remains hidden. Furthermore, this principle also ensures that the distributed models are discarded after they are combined with the larger, global model [4].

2.2 Technical Architecture of Federated Machine Learning

FML as a concept generally consists of data, users and the systems that combine, to operate the algorithmic portions. Within the FML framework, data is generated locally from devices, and then distributed across multiple repositories. These repositories are then used to build local models that are used to train their own devices. Which are then integrated back into the global federated model. This is done via the aggregator server, however there are some instances in which a decentralised approach is taken and a lead server is not required, this is seen above [16] (Fig. 2).

Taking a closer look FML, we can see there it is comprised of several different layers. These layers are the service layer, operator layer, infrastructure layer, cross-layer, algorithm layer [16]. The modules within the layers can be grouped together, as well as being hierarchical in nature. It should be noted that each of the modules and layers can interact with each other via the cross-layer. Above is a depiction of the different layers and modules that are found within the architectural framework of a standard FML algorithm [16].

As seen above, the CSF framework as a concept is composed of 5 layers, each with several modules within each layer. These modules interact within on another, often by utilising the Cross Layer as the go-between.

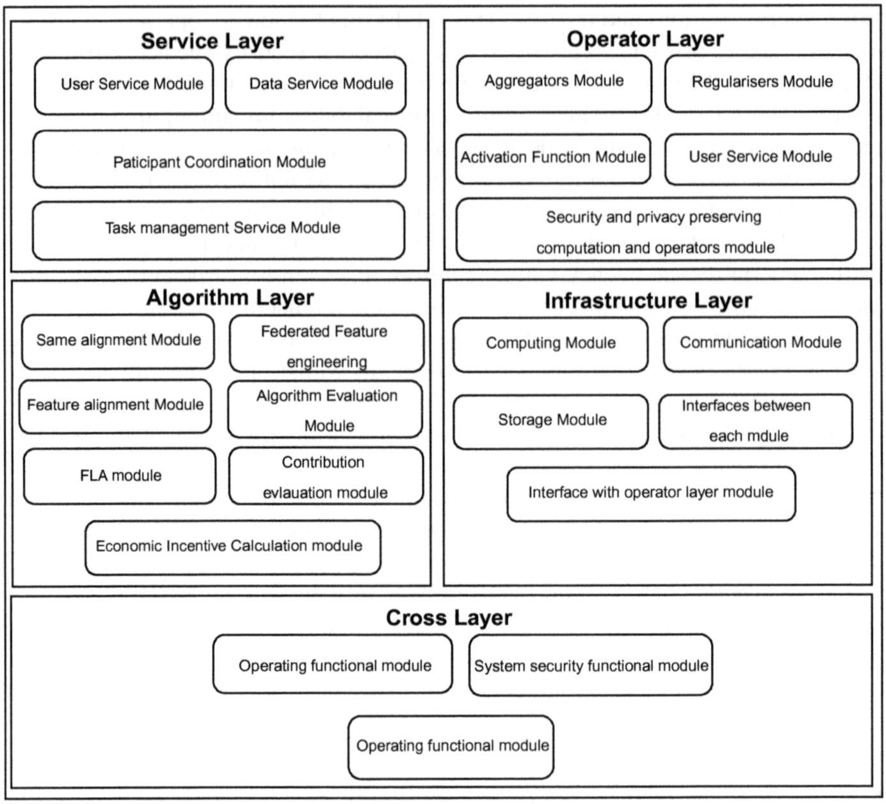

Fig. 2 Architectural framework of federated machine learning

2.3 Service Layer

The service layer implements logic based on requirements, and provides the services needed to federate machine learning for users. The service layer includes the user service module, the participant coordination module, data service module and the task management service module. The User Service Module supports the FML operations by enabling the many different components within the CSF. These components include but are not limited to a user interface (GUI, command line, or API) and task submission function. The participant coordination module enables the activity management from within the CSF. This involves CSF member management functions (the ability to monitor real-time status of FML systems) error tracking within the CSF and security management (secure data management functions). The data service module enables the management of the clients local data repository. The data service module also includes a set of import and export tools that allow for the importation and exportation of data and other models as well as a function that facilitates data owners to contribute their data to the assigned tasks, a metadata management scheme, a data contribution evaluation scheme, and finally, a module that judges the cost and benefits of data ownership, relating to each user and their role. The Task Management Service Module is responsible for managing FML tasks including modeling as well as including functions that enable it to submit functions that grant the user to ability to create and submit FML tasks, a query function that investigates learning tasks and logs, an error handling function that deals with abnormal terminations and a recovery function that assists with the task recovery [3].

2.4 Operator Layer

The Operator Layer provides the basic operations needed by the FML algorithm to be used on the large variety of learning algorithms. The operator layer further enables developers to implement a FML based on these aforementioned operators. The modules within this layer include an aggregator, activation, regularisation, optimisation module, as well as a computation operator. The aggregators module supports aggregation methods as well as enabling the customisation of each algorithm, with respect to denoted method of the aggregation, it should also be noted that the aggregators are assumed to still follow the assigned privacy-preserving principles as stated in the standard FML CSF (IE. Differential Privacy). The activation module includes functions from adopted methods based on traditional machine algorithms. Furthermore, the activation module operates by activating some federated functions, an example of this is Taylors Expansion Sigmoid. The regularisation module is used to ensure that machine learning functions are implemented correctly, and not underfitted or overfitted as well as also enabling the minimisation of loss from calculations. The optimisation module allows for the common implementation of methods for that fine tune machine learning methods. This module contains loss functions, optimisers and gradient processors for both traditional and federated machine learning algorithms. The computation operator is a module that ensures the basic principles of secure and privacy preserving computations are adhered to, these include but are not limited to homomorphic encryption, RSA encryption and differential privacy. This is done by enabling encryption and decryption of the participant data, by other users, as well as removing information related to the original data [3].

2.5 Infrastructure Layer

The infrastructure layer supports all the functions required by standard machine learning algorithms. The infrastructure layer provides capabilities of computing, storage, and communication of long-bit units related to the operator layer of the FML framework. The infrastructure layer also allows for interfacing between the operator layer, and each other component, these will be detailed below. The infrastructure layer has a computing component for computing long-byte data. It provides a clear FML task management scheme that monitors the running of the FML activities. The computing component also provides the API's for developing FML with regards to computing aspects, such as providing handles to computing resources and FML algorithms to reference to. The storage components primary function is to store operations within the FML framework, such as importing and exporting long-byte data from other storage components. A communication component for communication between long-byte data units, as well as working in tandem with the aforementioned FML task management scheme to enable the storage of said scheme. The storage component also enables the storage of each components access for APIs. The communications component's primary function is to provide communication with long-byte data, within the FML framework, as well as providing API's with the abilities of FML algorithm development. Furthermore, it also allows for the FML task management scheme to further communicate with the running machine learning algorithms. The infrastructure layer also interfaces between each component have 2 modules that facilitate the communication of both short and long byte data to and from other participants within the FML framework. The interfaces with the operator layer transmit the metadata and parameters of the FML learning tasks, requested from their operators. This function also gathers feedback from each component for their tasks [3].

2.6 *Cross-Layer*

The cross-layer module's function is to interact with the other layers (service, operator, infrastructure and algorithm) in order to support their capabilities. These include but are not limited to operating function, system security, monitoring and evaluation functions. The operating functional module has several abilities, these include but are not limited to a service catalog that lists all of its services for the FML system, a strategy management function that provides definitions, updates, strategies and other associated strategies for FML services. Furthermore, it also has an exception and problem management function that provides abilities to capture incident reports, as well as a delivery management function that ensures delivery of a functional interface to the user. A system security functional module will mainly provide guarantees of attributes of security (confidentiality, integrity, availability and privacy protection) needed for each functional layer, layer interaction and participant. This module also has an account management function, an authentication function, an authorisation and security policy function, a data integrity function, a data deletion function and a privacy disclosure function. The regulation and audit modules are responsible for making services regulate-able and audit-able to avoid privacy exposure. This module operates as the authority to determine whether the training, prediction and final model and other process's meet their requirements. These requirements relate to governance rules, storing audit information and the ability for real-time supervision of the module and the CSF contents.

2.7 Algorithm Layer

The algorithm layer implements the FML logic as well as simultaneously providing support for the service layer. This layer includes the sample alignment module, feature alignment module, federated feature module, federated machine learning algorithm module, algorithm evaluation module, contribution evaluation module and economic incentive calculation module. The sample alignment module identifies the overlapped samples of from different data and the associated sources, it should also be noted that this method does not disclose the information of the sample features. Furthermore, this module includes both input and output interfaces. The feature alignment module identifies overlapped features from different data sources, similar to the sample feature models, it does not discuss the non-overlapping features and other sample ID's. The module includes an input and output interface that lists all the feature names. The federated feature engineering module identifies overlapped input features and determines newer, related features that are relevant to the learning task. This module also contains an input section that utilises the selected dataset for feature engineering, as well as a listing of overlapping features and sample ID's. Furthermore, it also includes an output section that displays an updated data source. The FML algorithm module covers all the basic algorithms needed to complete FML for different scenarios or tasks. This module includes but is not limited to the algorithms that are related to tree-based, deep learning and semi/unsupervised machine learning operations. This module also includes a graphical user interface that displays both the input and out, with both including the listed data sources, features and samples. The algorithm evaluation module operates by evaluating the FML model in accordance to the specified evaluation measures stated within the denoted CSF. These evaluations include but are not limited to model performance, efficiency, privacy-preserving and security. The contribution evaluation module evaluates the contribution to the overall performance. It also has an interface that includes the training set data, testing data, as well as listing the overlapped features and samples from within the allocated data sources.

3 Operation of Federated Machine Learning Algorithm

FML as as paradigm operates on the premise of exchanging protected, private portions of the data and each site, then the trained model resides at the one node and or shared amongst other multiple clients. The main strands of FML can be broken down into 3 different types, horizontal, vertical and transfer learning [16].

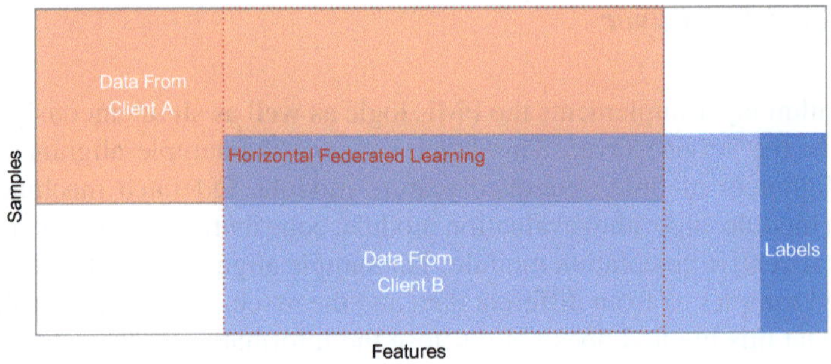

Fig. 3 Graphical depiction of HFL

3.1 Horizontal Learning

Horizontal Federated Learning (HFL) is also known as sample-partitioned or example-partitioned federated learning. This strand of FML can be applied to different scenarios in which data sets at different sites have overlapping features but differ in sample size. An example of HFL (or horizontal partition) can be viewed through the lens of a small rural bank. These smaller banks may have different users or clients from their respective areas and the total overlap of data is small in comparison but share a similar business model. The formal notation of HFL is noted as

$$Xi = Xj, Yi = Yj, Ii \neq Ij, \quad \forall Di, Dj, i \neq j. \tag{1}$$

Horizontal Federated Learning Equation

This notation states where the data feature space and label pair of both clients data sets (*Xi, Yi*) *1* and *2* respectively are assumed to be the same, as opposed to the user identifiers of (*Ii, Ji*) are assumed to be different. (*Di, Dj*) Denotes the data set of each party (*i*th and *j*th iteration) [14]. The typical operation of a client-server HFL system (also known as the master-worker system) operates by enabling *K* clients (or participants) with the same data structure, train a machine learning (ML) model with the assistance of the aggregator server. This server acts as the aggregator and coordinates the ML algorithm. This can be broken down into 4 steps. Step 1 consists of participants locally computing the training gradients, masking and then encryption said gradients with privacy techniques (such as differential privacy and secret sharing), the results are then sent to the server. Step 2 states that the master server performs the secure aggregation, an example of this is taking the weighted average [14] (Fig. 3).

The concept of HFL is best seen above. This diagram graphically demonstrates the equation of HFL. It shows the overlap between both sets of data. This is compared to Vertical Federated Learning (VFL) which is shown below.

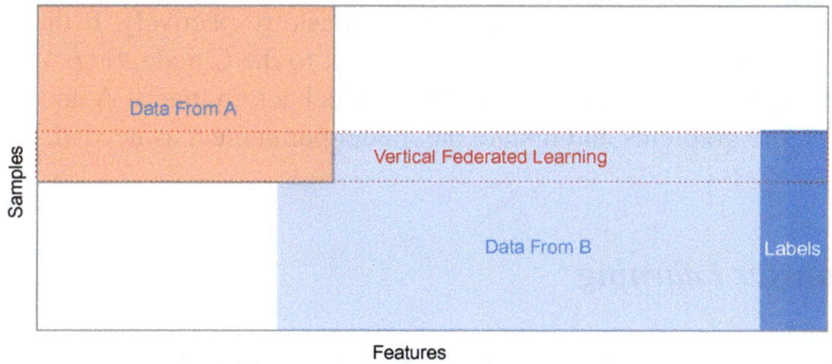

Fig. 4 Graphical depiction of VFL

3.2 *Vertical Learning*

Similarly, to HFL, Vertical Federated Learning (VFL) is used on data sets often gathered from businesses that have different goals and therefore different features. VFL (also known as feature-partitioned federated learning) can utilise heterogeneous features maintained by those organisations in order to build stronger ML models. Under the VFL framework, it is represented mathematically as

$$Xi \neq Xj, Yi \neq Yj, Ii = Ij, \quad \forall Di, Dj, i \neq j. \tag{2}$$

Vertical Federated Learning Equation

X and *Y* represents the feature space and label space. *I* represents the ID space, and matrix D represents the data that is held by the different clients and parties. The architecture of VFL differs slightly from HFL, due in large to the different types of participants. As stated above, let us assume that businesses A and B are both interested in operating a machine learning algorithm (MLA) jointly, with each having its own sensitive data. Furthermore, Business B has also labeled its data that the joint model needs to perform the required prediction tasks. As businesses A and B cannot exchange data directly due to privacy concerns, during the training process, a third-party collaborator (C) can be used. A trusted third-party can be implemented by a government organisation, or even a secure node (Intel Software Guard Extension) (Fig. 4).

VFL can be broken down into 2 parts, Encrypted entity alignment and Encrypted model training. The first portion, encrypted entity alignment refers to the idea that as user groups of two companies A and B are different, the system employs an encryption-based user ID alignment technique that confirms the identities to one another, without exposing each other's data. Part 2, also known as encrypted model training, can be further broken down into 6 steps. The first step revolves around C creating an encryption pair and sending the public key to A and B. Step two has A and B encrypting and exchanging the intermediate results for loss and gradient calculations. Step three includes Businesses A and B computing the encrypted gradients and subsequently adding an additional mask, respectively. B then computes the encrypted loss, A and B send the results back to the C node. Step 4 requires the decryption of gradients and loss sends the results back to that of A and B. A and B also unmask the gradients and update the model parameters as needed.

3.3 Transfer Learning

Federated Transfer Learning (FTL) is different from both HFL and VFL due in part to the idea that there are often situations in which there is no crossover of either features or sample spaces among the participants. However, FTL seeks to solve this issue by creating a means to build and operate an FML model with a minimal overlap of data as depicted below.

In one way or another, HFL and VFL require all participants to share features or other sample space, in order to build an effective model. If this cannot be done, FTL subsequently provides a solution for cross-domain knowledge transfer. This is evident, as there might only be a handful of labeled data which means that ML created modelling cannot be built reliably. FTL seeks to find the link between a resource-rich domain and a resource-poor domain and utilise the invariant dataset to transfer between both previously mentioned domains. There are several different approaches involved within FTL, these are instance-based, feature-based, and mode-based FTL, and can be applied to traditional HFL and FTL paradigms.

Instance-based FTL involves assisting the participating parties, allowing them to selectively choose the best features and samples, thus avoiding a negative effect on the model or the gradient. Feature-based FTL allows the participating parties to jointly learn a common feature representation space, in which distribution and difference from data sets can be relived and the knowledge can be transferred across different domains. Model-based FTL allows for the participants to collaboratively learn shared models that can benefit transfer-learning paradigms. An example of this would be learning to utilise pre-trained models or on parts of initial models for FTL tasks. Once the training samples have been combined with model based FTL tasks, this then allows for a more accurate computation of results. FTL, seeks to provide solutions for these scenarios. FTL is mathematically noted as

$$Xi \neq Xj, Yi \neq Yj, Ii \neq Ij, \quad \forall Di, Dj, i \neq j. \tag{3}$$

Federated Transfer Learning Equation

Within the above example, *Xi* and *Yi* shows the feature space and the label of space of the *i*th party. Furthermore, *Li* stands for the sample space, matrix *Di* represents the data set being held by that party.

FTL operates by learning from a sliver of cross-domain features and samples from organisations *A* and *B*, this is then extrapolated to be used for both parties to predict labels for the unlabeled dataset samples of organisation B [16].

4 Client Selection in Federated Machine Learning

In the original workings of FML, client selection as a concept was recommended as a way to reduce the cost and time needed for the global training round. The overall approach for the client selections is to enable the federated learning model owner to elicit thoughtful decisions as to which clients should be selected, then subsequently how to distribute the training tasks amongst them [11].

In the beginning, there was no original method for selection, and it was essentially just a random sample of clients. Later, a method known as FedCs was proposed by Nishio and Yonetani [8]. Nisio and Yonetani state that FedCs is a protocol that aims to work with clients with heterogeneous resources. This collaboration is done with mobile edge computing devices (Internet of things devices, smartphones and connected vehicles) in order to mathematically select the best placed client to complete the required computations [8] (Fig. 5).

Nishio and Yonetani [8] go onto describe the steps of operation for FedCs, these are depicted above. The first step of FedCs (initialisation) involves the clients informing the Mobile Edge Computing (MEC) operator of their information, this includes but is not limited to wireless channel states, computational capacities and size of data resources relevant to the current training round and task. Once this information is provided, the operator reuses this information in the second step (client selection), which is done in order to estimate the time required for the distribution and scheduled update and upload steps, and to direct the clients to these specific steps. The client selection step is integral as it allows the server aggregate as many client updates as possible within a specified time frame. The criterion that decides this is mentioned above, and at the same time this is being actioned, the operator schedules when the divided

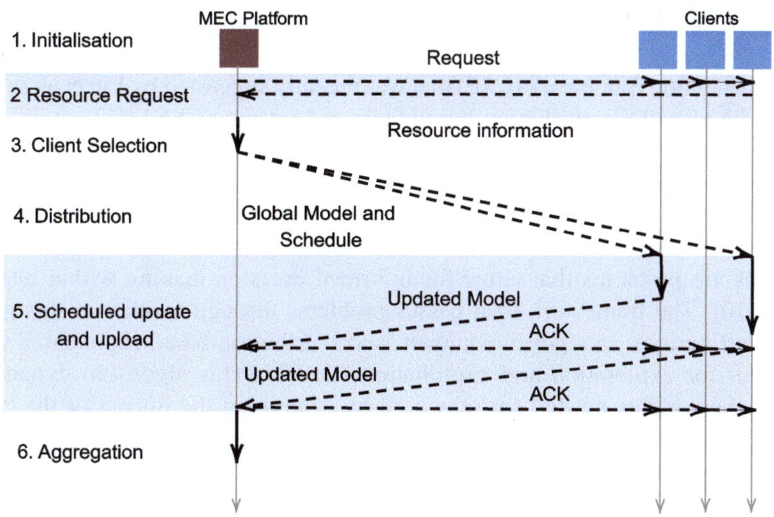

Fig. 5 The structure of the proposed framework for FedCs

up resources (also known as resource blocks or RBs) needed for model uploads are allocated to the selected clients. This is done in order to prevent bandwidth congestion on the MEC devices. Next, the distribution step ensures that the global model is transmitted to the specified clients via multi-cast protocols. The scheduled update and upload step ensures that the selected clients update their models in parallel and upload new parameters to the server using as per the amount of resource blocks denoted by the MEC allocator. After this occurs, the server then aggregates client updates via the standard federated learning method and then measures the model performance via examining the available validation data [8]. Although FedCs does solve one problem, with the issue of heterogeneity of device resource usage being a key one, it still falls short in some aspects. One aspect is that the authors never specify what clients are, they just state that they have set k = 1000. This means that we cannot be sure that the claim of FedCs being able to handle the heterogeneity of multiple devices is unable to validated. For the claim to be upheld, it would be beneficiary to state what devices are being simulated, as it is not clear just how varied the clients are [11].

Since this initial attempt at optimising client selection and then the subsequent suggestion of FedCs, there have been several more suggestions made, in order to find the best possible solution for selecting the strongest clients. The motivations for the continuing suggestions of strong CSFs stem from the notion of continual optimisation. As the adoption of internet of things, and other decentralised devices occurs, so will the increase in the size and geographical distribution of these networked clients. FML needs to be able to handle the subsequent increases, and as such can only be completed on an agile and non-bloated algorithm. These other suggested CSFs will be discussed below.

5 Client Selection Frameworks

A client selection framework (CSF) that was recently suggested by Lin et al. revolves around the idea of a contribution-based federated learning CSF [7].

The main contributions of this paper include a new client evaluation criterion which is based on contribution. The client's contribution is generated from the impact of each client on global mode accuracy. This paper also discusses transforming client selection into adversarial multi-armed bandit learning problems. Multi-armed bandit problems are problems that allow for informed decision making within uncertain events [10]. The framework then passes problems through another of this paper's main contributions, an algorithm known as a contribution-based exponential weight algorithm for exploration and exploitation (CBE3). This algorithm dynamically updates thè weights needed for selection, by examining the impact of the client's data. The third and final contribution of this paper are the creation of contribution factors that can be adjusted depending on the specific scenario needs [7].

Examining the first criterion, the selected criterion is based on the contribution of one client, versus another. The contribution is measured firstly by looking at the improvement of a single client to the global model's accuracy during that specific

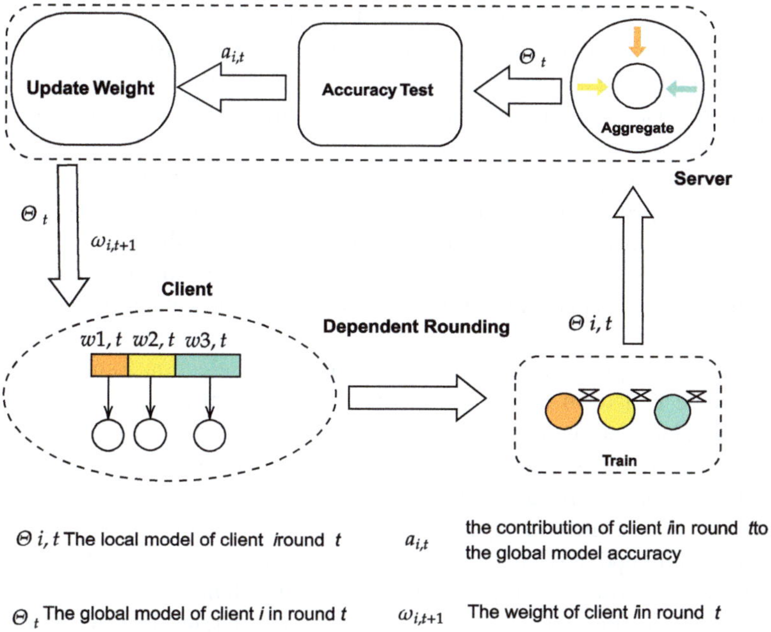

Fig. 6 The structure of the proposed framework for CBE3

communication round, the amount of improvement that the client generates on the model, is then used to describe the contribution. The second contribution (the CBE3 algorithm) uses this criterion to base its estimations on, which informs the algorithm as to which client is selected.

The CBE3 algorithm operates by first aiming to minimise the number of communication rounds required for the global model to reach a certain accuracy. The CBE3's goal is to select the client that maximises the amount of contribution per round, to accelerate the model's convergence. It does this by initialising the parameters of the global model and the selection weight of all clients to 1, this is done to ensure that all clients have the same probability of selection. The next process begins with determining the client's selection probability based on the computed weights. Next, based on those weights, The firs step of this portion of the CSF involves the clients, and their selection. The clients are selected at random to train local models of the clients and aggregate the global model, simultaneously recording the contributions of the client. Finally, updated selection weights are created and the probability of selection is assigned based on the unbiased estimation of the client's contributions from the current round. The process is iterated until training is no longer required. After each aggregation of the clients local model, the aggregated global model needs to be tested and the accuracy derived is then used in the calculation of the clients contribution [7]. This is seen in Fig. 6.

This algorithm was benchmarked against Greedy, K-Center and Random algorithms on the CIFAR-10 dataset. The results indicate that the CBE3 has strong adaptability, it also has a better balance between global model accuracy and convergence speed in some instances. This is contrasted to the Greedy and K-Center algorithms, which loses accuracy when convergence speed is sped up. One criticism of this paper is similar to previous paper. This paper does not specify the types of clients that are chosen and the specifications of those clients. This is important because one of the metrics of the paper refers to the speed of convergence, and this can be affected by the size and type of simulated client. What if the clients are simulated at an unrealistically small in size? This could impact your measurements of speed and obfuscate some of the results [7].

Zhao et al. [17] proposed an energy efficient client selection framework named FedNorm. The FedNorm framework (including FedNorm-E) seeks to find clients that provide a large amount of information in each round of FL training. Moreover, Zhao et al. also go onto suggest a more energy-efficient alternative that requires only the client selection step occur every certain round as specified [17].

FedNorm as a framework operates choosing the contributors that can provide significant weight updates. This is done by using the L2-Norm of local weight divergence during each round of training, this is done to identify the significance of a client. The flow of the algorithm starts of with the federated learning (FL) server choosing a candidate client from all the available clients to participate in the first round of training. Second, the clients compute local weight changes. At each round, the FL server sends its current global weight to the set of specified, chosen clients within that round. Each client found within that set computes the global model, its weight and the subsequent changes. After, this information is sent back to the server in conjunction with the local weight of the model. The CSF chooses nodes from the list that have the biggest weight change. Some of the drawbacks of the iteration of the FedNorm CSF is that each client is required to participate at all rounds with the client selection procedure. Then, [17] set out to rectify this. The authors found that only a small number of clients are used within the whole training process. FedNorm-E was then subsequently created due to the need to maximise energy efficiency. It means that only a small section of clients are used frequently throughout the training process. FedNorm-E operates similarly to FedNorm except it has a select clients procedure that includes a cache mechanism. In the first round of client selection, the algorithm chooses the top-max clients that have the largest weight change and then the largest weight change is chosen from that subset. This calculation may occur every specific number of rounds [17].

FedNorm as a whole CSF was found to be a more energy efficient CSF then other standard CSF's. Furthermore, [17] implemented an experiment in PyTorch and evaluated the proposed algorithms with FEM-NIST based datasets, with each dataset having both IID and non-IID data and data splits. These results demonstrated that the proposed algorithms out performs existing client selection methods (random selection, FedAvg and loss-based client selection) whilst reducing energy cost by participating clients numbers [17].

One issue found within this paper is how they measure energy efficiency. Zhao et al. [17] measure energy efficiency by suggesting that less clients needed during the operation of the algorithm is more energy efficient. As an general idea, that is correct, however, this paper has not actually measured the energy output, they have only stated that less clients will be used. As we know that this algorithm will be needed for a large quantity of heterogeneous devices, each of these could have different energy costs. This means that although they used less energy from less clients, they could still theoretically not save as much energy as claimed. Furthermore, this paper also never stated the energy requirements of the clients either. How can we be sure that this CSF is as energy efficient as it claims to be, if we don't know the measurements of energy efficiency.

Pang et al. proposed a CSF that incentivises the participation of heterogeneous devices within the federated learning paradigm [9].

This framework operates on the assumption that this is a typical FL scenario that involves a cloud server and a set of clients as well as the standard updating, training and modelling procedures. How [9] CSF differs is that it seeks to formulate a social cost minimisation problem in FL as an integer linear problem (ILP). This CSF also differs from existing frameworks as it determines not just the winner (clients selected for training) but the scheduling of them, as well as the total number of global iterations. This is done first by calculating the range of total global iterations needed, after this is done, a winner determination problem (WDP). To optimise, the WDPs are decomposed to solve computation time, and then bids are placed in order to bid for each WDP. The winner of the auction is rewarded with the task for training the model. The schedule for the winners is determined once again by using WDP and reformulating them to new ILP's. These new ILPs are solved by a greedy algorithm, and then iteratively selected as the client's schedule which has the lowest cost to the client [9].

This algorithm was benchmarked against FedAvg, Greedy Approximation Algorithm and AOnline. Overall, it can be seen that Pang et al.'s [9] proposed CSF was computationally efficient and achieves all the goals that were originally set out to reach. However, there are some issues. One issue was with how the algorithm was tested, like other papers, there was a lack of specification when it comes to the testing of said algorithm and CSF. It appears they have only tested this on a theoretical basis, they haven't attempted to apply it to a real-world scenario. This is evident as this paper is based purely on experimentation and has not incorporated a realistic look at the algorithm. What happens if the clients lose power during the bid process? This could happen in the real world, and yet, was not tested within this paper.

Li et al. suggested a CSF that is described as a novel self-adaptive federated learning framework in heterogeneous systems. This CSF is named FedSAE, which has access to the complete information of devices' past and present training tasks in order to predict what training loads are best suited for which device. Furthermore, this CSF also combines active learning to self-adaptively select participants. The framework aids in accelerating the convergence of the global model [5] (Fig. 7).

The proposed FedSAE CSF operates firstly by leveraging a workload prediction strategy based on previous workloads in order to predict the varied and affordable

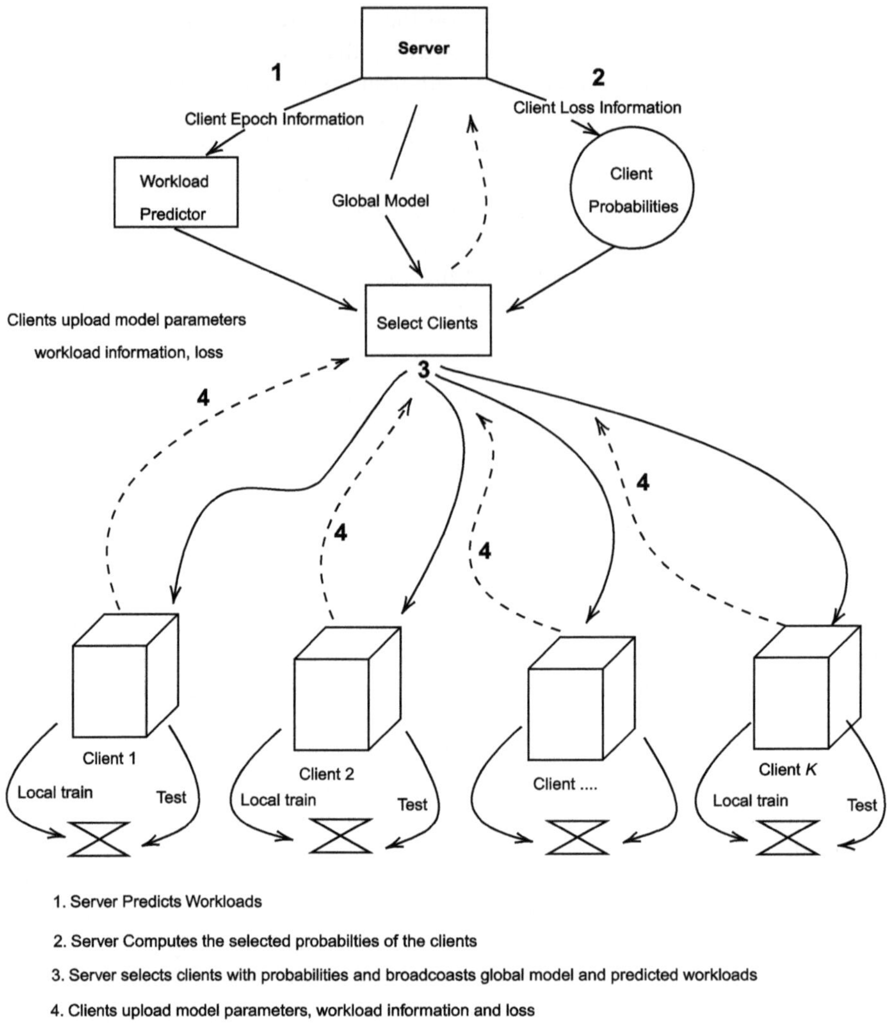

Fig. 7 The structure of the proposed framework for FedSAE

associated clients and the related tasks. With regards to the training strategy, FedSAE is closely related on FedAvg, with both selecting clients, broadcasting models, locally utilising stochastic-gradient descent, uploading local models and then aggregating the specific models.

Looking closely at the training process for FedSAE, this is best broken down into 4 steps, prediction step, conversion step, server participant selection step and model return step. The prediction step and the conversion step are integral in the operation of this CSF, as such, the authors go into further detail of these. The first step, (prediction step) relies on FedSAE to accurately predict the affordable and accurate workloads of their historically completed tasks. The first step underpins the suggested algorithm as, for the clients to participate, they must be able to self-adaptively adjust, if they do not, they are more likely to drop out. Therefore, it can be seen that it is necessary to predict the affordable workload of clients in a sufficient and correct manner. FedSAE manages this prediction by limiting the maximum and minimum workloads. Each client will maintain a pair of these workloads and then increases its workload until it drops out or completes the difficult workload. The client completes the smaller (easier) workload first and then the harder (bigger) one second. The second step (conversion step) occurs when the server converts the clients training loss into a selected probability. The conversion step operates by firstly changing the cross-entropy loss of each client into the value of each training measure, in order to measure the importance of the client. Then, the server converts the value into a selected probability, and then selects the participants for each round according to the created probabilities. The third step (server participant) ensures that the correct participants are chosen based on the results from the conversion step, and then broadcasts the true global model and the predicted model of the chosen participants to said, participants. Lastly, the fourth step (model return) returns the model parameters, information needed for newer task completion and the next round of learning and training [5].

One issue that needs to be pointed out is to do with the basis of this algorithm and its use of prediction. As this algorithm is based on the history of the device, how this history is calculated is of great importance. It is seen that if client has no other history, then a pair of numbers is chosen, based on the lowest and highest possible s needed to complete the algorithmic computations. This is essentially utilising the process of elimination.

It should be noted that in recent years, there has been an influx of new suggestions regarding CSFs. This is seen in Tables 1 and 2, where different suggestions have been made in order to attempt to address the issue of client selection within small devices. These suggestions vary greatly, going from the aforementioned FedCs and using a resource check, to using FTL and creating another model in order to further aid global model creation.

This naturally leads to our proposed framework. Having performed an indepth literature view of the current frameworks, and even suggesting several newer ones, we will now discuss what makes a good CSF, and then how our proposed framework fits within this criteria.

6 Our Proposed Framework

To suggest a stronger and more efficient CSF, we must look at the aforementioned issues with Federated Machine learning, and then address these. These issues include but are not limited to the often large amount of data created by internet of things (IoT) devices and the subsequent large computational power needed to operate on dense data silos. Data ownership and privacy is another is another issue that should be

Table 1 Different client selection frameworks within FML

Framework	Publication date	Dataset	Motivation	Methodology
FedCs	2019	CIFAR and MNIST	Optimise FML operations on MEC framework	Clients receive an offer and distribute their information which determines the clients computations
FedSAE	2021	FEMNIST, MNIST, Sent140, Synthetic(1,1)	Optimise client selection based on the previous history of learning tasks	Framework selects the client device with the biggest value to the global model
FedNorm	2022	FEMNIST based datasets (IID and non IID)	To optimise client selection framework focusing on heterogeneity	Selects clients to finish training tasks based on how much they provide to the global model
CBE3	2022	CIFAR-10	Seeks to improve global model accuracy whilst sticking to computational boundaries	Transforming client selection into MAB problems. The suggested algorithm allocates selections based on these contributions
AFL [9]	2022	MNIST, Fashion-MNIST, CIFAR-10	Suggests an mechanism that encourages participation from heterogeneous clients	Decomposes into several problems. Then suggests winners and rewards these clients

addressed. This includes data integrity, as well as authentication of devices within the network. Perhaps the most pressing issue with regards to the current CSF paradigm is that of heterogeneity of devices. Heterogeneity and the related issues of interoperability between devices needs to be addressed by a flexible CSF. Scalability is also another issue that a current generation CSF must also address, with the uptake of small smart devices increasing rapidly, this therefore means the size of the ecosystem is increasing in size, making it largely unstable. This instability can manifest in different ways such as loss of connection of slower uploads or modelling speeds. The instability could also cause unbalanced numbers of data samples, which can therefore disproportionately affect the training rounds and subsequent models gained from these [16].

Now that we know what the biggest issues within FML are, we can start to suggest what would make a strong CSF for federated machine learning. We can see that this CSF must be able to operate within different device makes and models. The CSF must

Table 2 Table of multiple client selection frameworks

First entry	Second entry	Third entry	Fourth entry	Methodology
FedAux [2]	2022	CIFAR-10, FMNIST	Adjust below average client selection schemes that don't account for feature-drift	Create an intermediate model and transfer knowledge to traditional clients
HPFL-CN [6]	2022	Air-India, U-Air	To enable the optimisation of edge servers to be modelled effectively	Clustering edge servers with similar distributions, and then executing a customised model for each cluster
HACCS [13]	2022	FEMNIST, CIFAR-10	Optimise client selection and for model convergence as well as ensuring the inclusion of different types of data	Identifies heterogeneity by representing all data distributions instead of each specific device
FedSec [1]	2022	MNIST, CIFAR-10 and FMNIST	Optimise nodes working on larger datasets	Allows faster node learners to train more steps which then expadites the central model

be able to handle the large amount of data generated, and subsequent varied local data sets. The CSF must adhere to all authentication and data handling guidelines, including but not limited to the GDPR as well as device anonymity and privacy. The CSF needs to be lightweight and scalable in nature, being able to make connections with other clients quickly, and be able to operative effectively. The CSF must allow for FML to take place with minimal to no downtime, this will then allow for accurate modelling and representation of the client devices and their impact on the global model. Our proposed framework is based on the idea that it should be able to be used by different devices and different manufacturers, with a better outcome then the current frameworks found within academia [16].

7 Applications for Federated Machine Learning Using Client Selection Frameworks

Federated machine learning has many different areas of application. Client selection can assist these areas with allowing for the optimisation of resource usage. As mentioned above, the increase in usage of IoT devices has increased the network size and scale and geographic dispersion, as such the need for optimisation has also increased.

The areas that FML could be applied to, include but are not limited to sectors in which data cannot be privately aggregated for training machine-learning models. This could include areas such as intellectual property rights, privacy protection and data security. Yang et al. [15] give a specific example regarding retail sales. Yang et al. [15] further discuss potential applications within this sector and state that FML could be used to for personalisation services, including product recommendation and sales services. These recommendations could be gleamed from different areas such as the users bank and social media accounts. Cross company cooperation could theoretically work utilising FML, however the hardest barrier of entry would be granting permissions to the clients for accessing the other clients accounts. An example of this would be granting permissions for a FML algorithm, to operate between the Commonwealth Bank and Woolworths supermarket store to provide real time marketing for purchases that the user can afford [15].

Yang et al. [15] also suggest another potential application for FML in the field of multiparty database queries. These queries could operate firstly, by not exposing the data to any outside. The example provided by the authors involves the banking industry, and states that FML could be a way to detect multi-bank lending. Multi-bank lending is a threat to financial security and stability. FML can solve this by first utilising an encryption feature, each bank could list their users and then take the intersection of the encrypted list. The decryption of this list would then show the good users to the other party.

Healthcare can also benefit from these implementations, with implementing FML creating whats known as smart healthcare. Smart Healthcare involves the usage of FML techniques, in combination with medical records relating to diseases, gene sequences as well as medical reports, could allow for a creation of a smarter, better machine learning models. As current machine learning models are outdated and have slowed progression in this sector, the new use of FML could allow for the training of larger data sets to be specifically used within the healthcare sector. Furthermore, FTL could also be used to label missing pieces of data. This would also expand the scale of available data, ensuring that models can be trained on the much larger data sets mentioned above [15].

Another important current usage for FML is that of its combination with blockchain technology. Blockchain technology is a decentralised, immutable ledger of records that empowers participated devices named miners to contribute to the network. Each miner keeps one copy of the ledger locally, whilst competing to win the chance to create a new block which contains more records of transactions [Wang and Hu]. Blockchain syncs well with FML as it allows the centralised nature of FML (with an aggregator, as mentioned above) to continue working, even if the server suffers has downtime. This means that the blockchain nodes would take charge of the aggregation, in the event of server failure. Another way in which this is blockchain assist with the FML paradigm is with verification. Blockchain provides verification mechanisms for FML ecosystems via authentication methods, and thusly remove unverified or malicious attacks or updates on the global or local models. Moreover, blockchain also helps with reward distribution (economical incentives such as cryptocurrency), and actively encourages participation positive behaviour within the FML ecosystem. The final way that the authors suggest that [12] suggest blockchain assists with learning data. Blockchain utilises assists with learning data can be stored on the distributed ledger. Once data is recorded on the ledger they cant be tampered and subsequently allow authorised clients can access particular ledgers to retrieve the data, thus helping to improve the training efficiency [12].

8 Conclusion and Future Work

There are several potential future research topics relating to the current client selection paradigm. The current CSF issues revolve are resource management, client number maximisation, client dropout and reliable client selection. Some specific issues found within these are suggested by [11]. Wahab et al. [11] suggest that there could be upwards of 12 current challenges that would be suitable for future research options. The first issue relates to studying the impact of wireless network conditions on the accuracy of the federated training step. The second issue is discovering a best approach to efficiently utilise the limited computation size and power of smaller devices, in order to maximise client training strength. Another area of potential research focuses on dynamically adapting the frequency of performing global model aggregation, in order to minimise the total amount of resources that clients need. The fourth challenge, involves researching resource heterogeneity across multiple client devices, and subsequently making efficient use of the different amounts of resources, in order to optimise training performance on the each node. The fifth challenge relates to the allowance of a partial peer to peer model, in which updates are shared and clients synchronise with said updates partially, in order to minimise the effect on performance. Next, the sixth challenge suggests that further research could be put into finding out the uncertainty and absence of previous knowledge and a history of the clients resource status and network connections. The seventh challenge relates to creating a method of evaluation in which each of clients' contributions are to the federated training cycle is rated, in order to improve future selections. The eight challenge involves assessing the reliability of the participating clients to further minimise the undesirable behaviours that could occur. Wahab et al. [11] describe the ninth challenge investigating mutual trust relationships among clients, and fostering distributed local model updates as well as the ability to share them. The tenth listed challenge is finding a way to protect the trust establishment process against malicious attackers that attempt to influence results. The eleventh challenge is investigating the trade-off between increasing client participation numbers and the subsequent drop in distribution of data and model update speeds. Lastly, the final challenge described by [11] states that a way to further research scheduling and selection could be to look at different factors such as data size, computational and network resource capabilities, and reliability in the client selection process [11].

Wahab et al. [11] also suggest further research directions, one such direction relates to discovering a way to reduce the frequency of carrying out global model aggregation, which therefore reduces the amount of resources needed on devices. This could theoretically be achieved by looking at FML through a dynamic lens, auto adapting the updating of the model to the needed frequency (based off of the desired accuracy). Another topic could be to design intelligent resource management strategies that enable the cross-linking of training loads, to each different clients with the amounts of available resources on that clients specific device. The authors do also point out that trust is an important aspect of FML, and trust between clients and the server are needed, inter-client trust is also needed, and is a potential research topic for future use. That is, being able to trust other clients would enable client model sharing updates between one another client node, as opposed to having to use the aggregator servers model [11].

References

1. Gao Z, Duan Y, Yang Y, Rui L, Zhao C (2022) FedSeC: a robust differential private federated learning framework in heterogeneous networks. IEEE, pp 1868–1873. https://doi.org/10.1109/wcnc51071.2022.9771929
2. Gu H, Guo B, Wang J, Sun W, Liu J, Liu S, Yu Z (2022) FedAux: an efficient framework for hybrid federated learning. IEEE, pp 195–200. https://doi.org/10.1109/icc45855.2022.9839129
3. IEEE (2021) IEEE guide for architectural framework and application of federated machine learning. IEEE Standard 3652, pp 1–69. https://doi.org/10.1109/IEEESTD.2021.9382202
4. Jatain D, Singh V, Dahiya N (2021) A contemplative perspective on federated machine learning: taxonomy, threats & vulnerability assessment and challenges. J King Saud Univ Comput Inf Sci 34(9):6681–6698. https://doi.org/10.1016/j.jksuci.2021.05.016
5. Li L, Duan M, Liu D, Zhang Y, Ren A, Chen X, Tan Y, Wang C (2021) FedSAE: a novel self-adaptive federated learning framework in heterogeneous systems. IEEE, pp 1–10. https://doi.org/10.1109/ijcnn52387.2021.9533876
6. Li Z, Chen Z, Wei X, Gao S, Ren C, Quek T (2022) HPFL-CN: communication-efficient hierarchical personalized federated edge learning via complex network feature clustering. IEEE. https://doi.org/10.1109/secon55815.2022.9918588
7. Lin W, Xu Y, Liu B, Li D, Huang T, Shi F (2022) Contribution-based federated learning client selection. Int J Intell Syst. https://doi.org/10.1002/int.22879
8. Nishio T, Yonetani R (2019) Client selection for federated learning with heterogeneous resources in mobile edge. IEEE, pp 1–7. https://doi.org/10.1109/icc.2019.8761315
9. Pang J, Yu J, Zhou R, Lui J (2022) An incentive auction for heterogeneous client selection in federated learning. IEEE Trans Mob Comput 1–17. https://doi.org/10.1109/tmc.2022.3182876
10. Patil V, Ghalme G, Nair V, Narahari Y (2021) Achieving fairness in the stochastic multi-armed bandit problem. J Mach Learn Res 22(174):1–31
11. Wahab O, Mourad A, Otrok H, Taleb T (2021) Federated machine learning: survey, multi-level classification, desirable criteria and future directions in communication and networking systems. IEEE Commun Surv Tutor 23(2):1342–1397. https://doi.org/10.1109/comst.2021.3058573
12. Wang Z, Hu Q (2021) Blockchain-based federated learning: a comprehensive survey. arXiv preprint arXiv:2110.02182
13. Wolfrath J, Sreekumar N, Kumar D, Wang Y, Chandra A (2022) HACCS: heterogeneity-aware clustered client selection for accelerated federated learning. IEEE. https://doi.org/10.1109/ipdps53621.2022.00100

14. Yang Q, Liu Y, Chen T, Tong Y (2019) Federated machine learning: concept and applications. ACM Trans Intell Syst Technol (TIST) 10(2):1–19
15. Yang Q, Liu Y, Chen T, Tong Y (2021) Federated machine learning. ACM Trans Intell Syst Technol 10(2):1–19. https://doi.org/10.1145/3298981
16. Yang Q, Liu Y, Cheng Y, Kang Y, Chen T, Yu H (2020) Federated learning. Springer International Publishing. https://doi.org/10.1007/978-3-031-01585-4
17. Zhao J, Feng Y, Chang X, Liu C (2022) Energy-efficient client selection in federated learning with heterogeneous data on edge. Peer-to-Peer Netw Appl 15(2):1139–1151. https://doi.org/10.1007/s12083-021-01254-8

Explainable Anomaly Detection in IoT Networks

Zahra Jadidi and Shantanu Pal

Abstract Due to the increasing number of threats against Cyber Physical System (CPS) networks, security monitoring in these networks is challenging. Machine learning methods have been widely used to analyse network data and detect intrusions automatically. However, these automated intrusion detection systems (IDSs) are black boxes, and there is no explanation for their decision. Therefore, explainable machine learning techniques can be used to explain the reasons behind the decision made by machine learning-based IDSs. However, there is no sufficient study on explainable IDSs in CPS networks. The other challenge in CPS networks is the growing volume of data. A NetFlow-based analysis is a scalable method suitable for a high volume of data. However, the efficiency of such a method in CPS networks has not been sufficiently investigated. In this chapter, we address these challenges by proposing an explainable NetFlow-based IDS (X-NFIDS) for CPS networks. The Internet of Things (IoT) environment is used as an example of CPS networks. To demonstrate the feasibility of our approach, we perform some preliminary studies of the proposed method using two NetFlow datasets for IoT.

Keywords Machine learning · Anomaly detection · IoT networks

1 Introduction

Machine Learning (ML) has been widely used in Cyber Physical Systems (CPS) security. CPS data, including network traffic and device logs, can be used to train ML models to detect abnormal behaviour. The increasing availability of large amounts of

Z. Jadidi
School of Information and Communication Technology, Griffith University, Gold Coast Campus, Brisbane, QLD 4222, Australia
e-mail: z.jadidi@griffith.edu.au

S. Pal (✉)
School of Information Technology, Deakin University, Geelong, VIC 3220, Australia
e-mail: shantanu.pal@deakin.edu.au

S. Pal et al. (eds.), *Emerging Smart Technologies for Critical Infrastructure*,
Smart Sensors, Measurement and Instrumentation 44,
https://doi.org/10.1007/978-3-031-29845-5_4

data increases the use of *NetFlow-based* anomaly detection. While NetFlow-based analysis has been proposed as a scalable anomaly detection method for high-speed networks, there is not sufficient research on the application of this method in CPS networks [5, 7].

Although modern ML technologies have significantly matured, there are still issues and doubts. While the accuracy and complexity of ML models are growing, there is no explanation for the reason behind their conclusions. ML models have been 'black-box' models because nobody can clearly explain data from input to output. Explainable ML methods have been developed to provide useful insights for users to comprehend ML working patterns, algorithms, learning processes, and results [13, 15]. In multiple popular areas, e.g., biotechnology, NLP (Natural Language Processing) and machine vision, model explanation and interpreting have been highly focused and well-developed. However, when it comes to cybersecurity, the disturbing seems not that ignorable [23].

In addition, even adopting just the ML method, ML systems are vulnerable to hostile attacks that limit the use of ML in general. Adversarial ML attacks are examples of attacks on ML models. If we can make the ML process explainable, the analysis would greatly benefit the development of the models' security level. In the end, the descriptive approach to the decision-making process of the black-box classifier is one of the necessary factors in successfully implementing these models.

ML-based intrusion detection systems (MIDSs) can detect malicious activities by learning the various patterns of network behaviour. An explainable IDS can provide local and global explanations. By combining local and global reports, the structure improves MIDS interpretation. The local description explains why the model makes certain decisions about specific inputs. The global description illustrates the main characteristics extracted from the IDS and the relationship between the values of the characteristics and the different types of attacks [12].

To address the challenges mentioned above, in this chapter we present a novel NetFlow-based intrusion detection for IoT networks. In particular, we propose an explainable NetFlow-based intrusion detection system (X-NFIDS) for an IoT network. A popular technique for model explanation is SHAP (SHapley Additive exPlanations) [4]. SHAP method is a game theoretic method used to explain the outputs of ML models. This technique is used to explain the predictions made by our NetFlow-based IDS. The major contributions of the chapter can be summarized as follows:

- Development of an explainable NetFlow-based IDS (X-NFIDS) in IoT networks.
- Implementation of a NetFlow-based IDS (NFIDS) in IoT networks.
- Evaluation of the proposed X-NFIDS using two real-world NetFlow datasets.

The rest of the chapter is organized as follows. In Sect. 2, we present related work. In Sect. 3, we discuss the proposed X-NFIDS for IoT networks. In Sect. 4, performance is discussed based on the results. Finally, in Sect. 5, we conclude the chapter.

2 Related Work

An IDS is a system designed to, as the name suggests, detect intrusions. In general, IDSs do so by examining evidence, e.g., network logs for unusual behaviour which may indicate malicious behaviour [6]. IDSs then alert the appropriate personnel, e.g., a system administrator, who will then make a decision based on the contents of this alert. Early applications of ML aimed to address this by learning so-called user 'profiles' by determining a typical pattern of behaviour for a user and thus better detecting anomalous actions [21]. Such a system is more adaptable and, by observing unusual behaviours, may detect attacks other than those it is specifically programmed for.

NetFlow is a Cisco protocol for network traffic collection. NetFlow-based anomaly detection is a scalable method for high-speed networks. NetFlow-based analysis extracts features from packet headers, and as it does not depend on packet payload, it can even analyse encrypted traffic. The features extracted by NetFlow provide key security features which are crucial to improve the performance of ML models in intrusion detection [2, 17]. However, while CPS networks generate a large amount of data and NetFlow-based analysis can help reduce the volume of data, there is insufficient research about the application of NetFlow-based anomaly detection in CPS networks. Only limited papers have studied this issue in CPS networks [7].

The issue with the existing NetFlow-based systems is that they are based on traditional ML methods. As stated earlier, the increasing complexity of ML models makes it challenging (or impossible) to explain the justification behind a decision. This is not useful for cybersecurity personnel as they can neither rule out false positives nor improve the security of existing networks. This is why many proposals with a theoretical 100% detection rate in testing have yet to be adopted in service [8]. Therefore the research gap addressed in this chapter is designing an explainable NetFlow-based IDS using an ML approach and a method of explainability.

Table 1 is a collection of relevant papers in XAI and IDS that have been collated and the method implementation. All of these papers use the SHAP method to explain the decision made by ML methods. Inspired by these papers, we propose a SHAP method in our explainable X-NFIDS for the IoT network. However, despite these papers, our detection method is explainable NetFlow intrusion detection.

3 Proposed X-NFIDS for IoT Networks

A SHAP-based framework is proposed in this chapter to provide an explainable NetFlow-based intrusion detection in IoT networks. The proposed framework is illustrated in Fig. 1. In this framework, the SHAP value is used to explain the prediction made by an ML classifier. After using the SHAP value for prediction, four results are related to the local and global explanations, respectively. The SHAP waterfall plot and decision plot are located in the local explanation, and the SHAP violin plot

Table 1 Comparison of available explainable intrusion detection systems

References	Machine learning method	XAI method	Datasets
[11]	Deep neural network	SHAP, LIME, CEM, ProtoDash, BRCG	NSL-KDD
[19]	Deep feed forward + random forest	SHAP	CSE-CIC-IDS2018, BoT-IoT, ToN-IoT
[25]	Random forest	SHAP	CIRA-CICDoHBrw-2020
[8]	Random forest	SHAP	NF-BoT-IoT-v2, NF-ToN-IoT-v2, IoTID20
[22]	Random forest	SHAP	CICIDS
[16]	Bootstrap aggregation + random forest + extra trees + XGBoost + Naive Bayes	SHAP	AWID
[14]	XGBoost	SHAP + LIME + ELI5	IoTID20
[23]	KNN + random forest + SVM + ResNet + DNN	LIME + SHAP	NSL-KDD
[1]	Convolutional LSTM encoder-decoder	SHAP	ADS-B training set

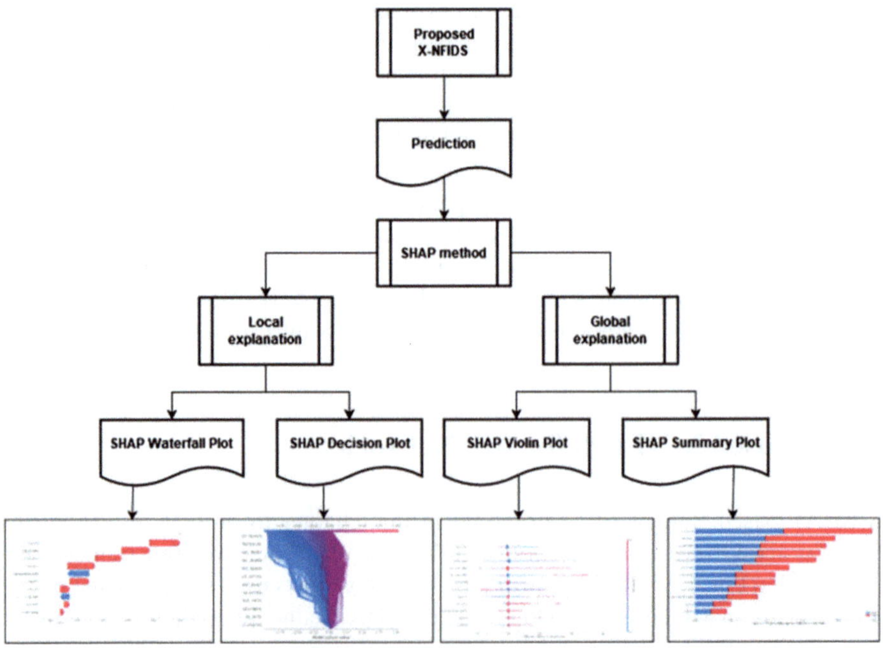

Fig. 1 The structure of the proposed framework for X-NFIDS

and SHAP summary plot are located in the global explanation. Then, it starts the implantation base on SHAP value features. Finally, we used two datasets to evaluate the performance of our X-NFIDS.

3.1 SHAP Methodology

SHAP explanations are a popular feature-attribution mechanism for explainable ML. The Shapley value was created by Shapley [20] in 1953 used to clarify the importance of network data features in the detection of network attacks. In this chapter, we use SHapley Additive exPlanations (SHAP) [10] to test the effect of specific outcome on the model. SHAP value is a common explainable ML technique developed by Lundberg and Lee [10]. It is based on an additive feature importance method of calculating the Shapley values, known as KernelSHAP and TreeSHAP. In our model, SHAP gives a value that can be important to each feature in the given prediction. These values are calculated for each prediction separately and do not cover a piece of general information about the entire model. High absolute SHAP values indicate high importance. However, values close to zero indicate the low importance of a feature. The SHAP explanation is effective in domain explanation [9].

In this chapter, the SHAP method in implementation is used to explain random forest data and generate different people's readable graphics that explain the training and the testing. Also, the SHAP method will generate two different types of graphs global explanation and local explanation [19]. Further, as a framework in our experiment, SHAP combines local and global explanations. The global explanation is associated with the average behaviour of the method after some aggregation. Local explanations explain how a model makes decisions. There is an important function that SHAP is a method used to interpret predictions based not only on the importance of features based on a complex training model but also on the predictions of the current test sample [24].

SHAP value explanation includes a waterfall plot and a decision plot for the local explanation and a violin plot and a prediction summary plot for the global explanation. In terms of explainable ML, global explanation means users can analyse the ML model from an overview of essential features. The advantage of global explanation using SHAP value is finding the positive and negative output in a whole ML mode. The local explanation involves how the user uses SHAP to make decisions, and it is through every SHAP value while using SHAP value to explain why ML model decisions [8].

3.2 Random Forest

It is an advanced decision tree. It is also a combination of many decision trees, adding randomly assigned training data to significantly increase the final result [3]. Also,

Table 2 Evaluation of X-NFIDS using two NetFlow datasets

Dataset	NF-BoT-IoT (%)	NF-ToN-IoT (%)
Accuracy	99.2	100
Recall	99.6	100
TPR	83.2	100

random forest is an overall learning algorithm that uses Bagging plus random feature sampling. On the other hand, random forest is based on ensemble learning methods, and ensemble learning method is for classification, regression and aggregating the result. The well-known classification method is bagging and boosting. In the bagging method, data samples in the training set are randomly selected and replaced, which means that a single data point can be selected multiple times. After generating a few data samples, the models are trained independently, depending on the type of task.

4 Results and Discussion

We employ two real-world NetFlow datasets to test the X-NFIDS in IoT networks, NF-BoT-IoT and NF-ToN-IoT [18]. NF-BoT-IoT has 600,100 samples, of which 97.69% attack samples and 2.31% benign samples. NF-ToN-IoT has about 1 million samples, out of which 80.4% are attack samples and 19.6% are benign samples. A random forest classifier is implemented to detect attacks in these datasets. Training and testing datasets were randomly selected from NF-BoT-IoT and NF-ToN-IoT.

From the random forest classifier prediction, in the NF-BoT-IoT dataset, there are accuracy of 99.2%, Recall 99.6% and True Positive Rate (TPR) 83.2% (Table 2). In the NF-NoT-IoT dataset, the accuracy, Recall and TPR are all 100%. These performance metrics are calculated as Eqs. 1, 2, and 3 where (TP) is true positive, (TN) is true negative, (FP) is false positive, and (FN) is false negative.

$$Accuracy = \frac{TP + TN}{TP + TN + FP + FN} \tag{1}$$

$$Recall = \frac{TP}{TP + FN} \tag{2}$$

$$TPR = \frac{TN}{TN + FP} \tag{3}$$

After training the random forest model, SHAP values were used to explain the predictions. Figures 2, 3, 4 and 5 are the plots provided by SHAP. The graphic in Fig. 2 is the decision plot of all the data types, which is the whole model output values. The vertical line presents important features and shows how to make the decision in each feature of the IDS model. The colored lines show the prediction.

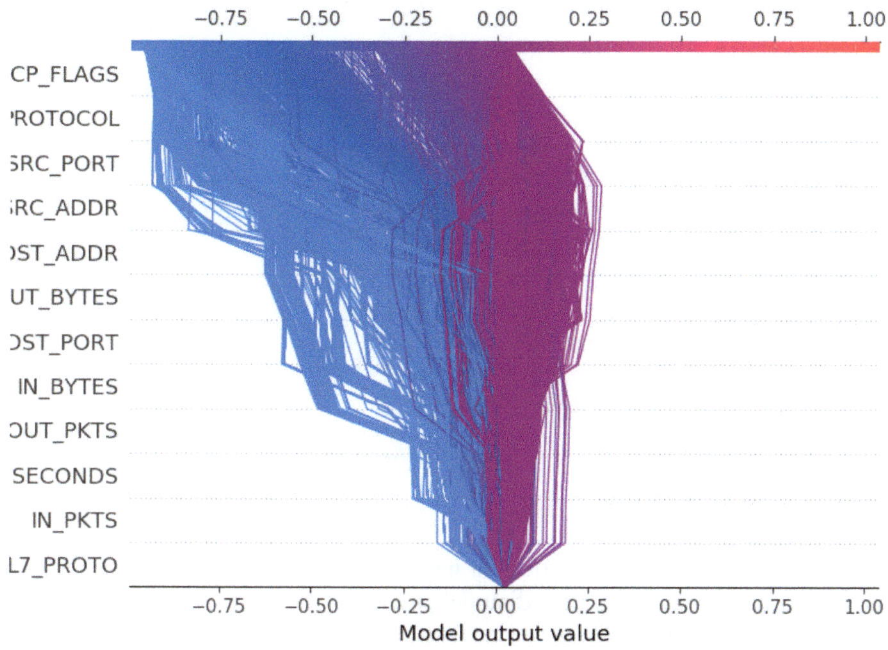

Fig. 2 Decision plot of X-NFIDS

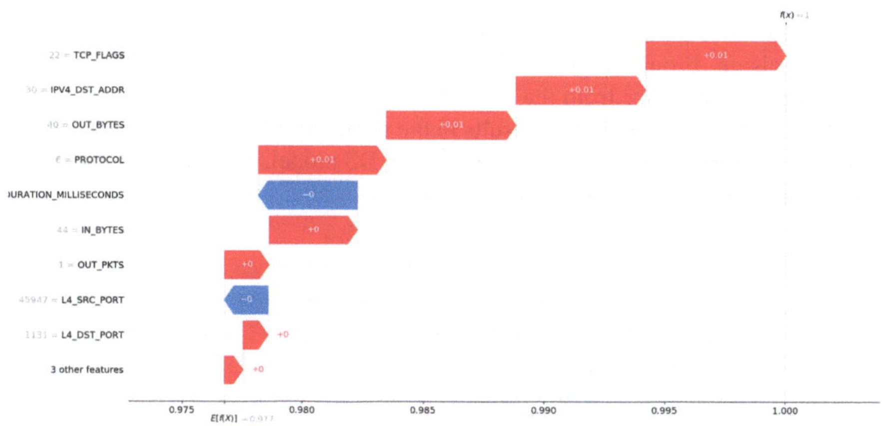

Fig. 3 Waterfall plots of X-NFIDS

Local interpretation of detail features involved in X-NFIDS decisions using waterfall plots is shown in Fig. 3. Where $f(x)$ is the log odds ratio of the predicted features, and $Ef(x)$ is the mean value of the log odds ratio of the predicted features. The horizontal line of a waterfall plot shows the expected values of the outputs. Then, the

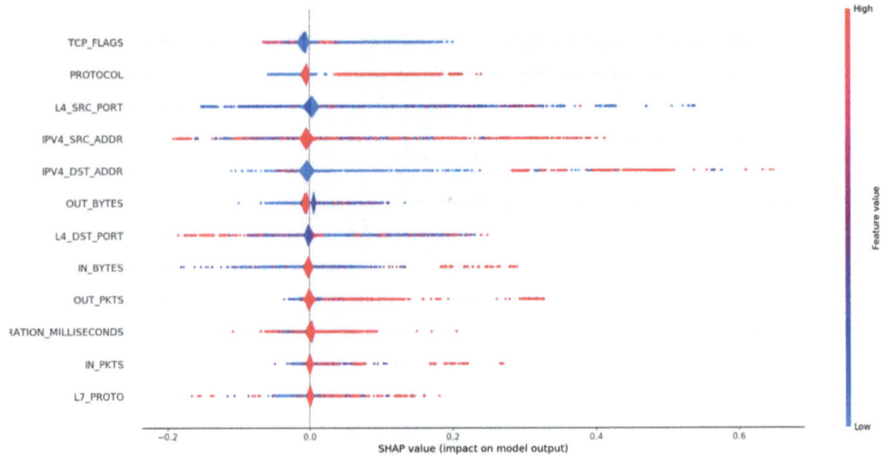

Fig. 4 Violin plot of X-NFIDS

vertical axis shows the positive (red) or negative (blue) contribution of each feature moves the value from the expected model prediction.

Figure 4 shows the results of important selected features using the SHAP model. Each row in the figure represents a detailed feature in the IDS model, and the horizontal information is SHAP values. The more redder colour means larger values. For example, this shows lower 'ipv4_src_addr' feature corresponds to a large value, which means that the lower component of 'ipv4_src_addr', indicates there are more in IDS decision. From the graph, the feature of 'ipv4_dst_addr', feature value is higher when the distance from the ground.

Figure 5 shows the prediction results in the summary plot, which is the summary of the development in each feature of the X-NFIDS model. The length of bars in this plot presents the influence of the feature on the classifier predictions.

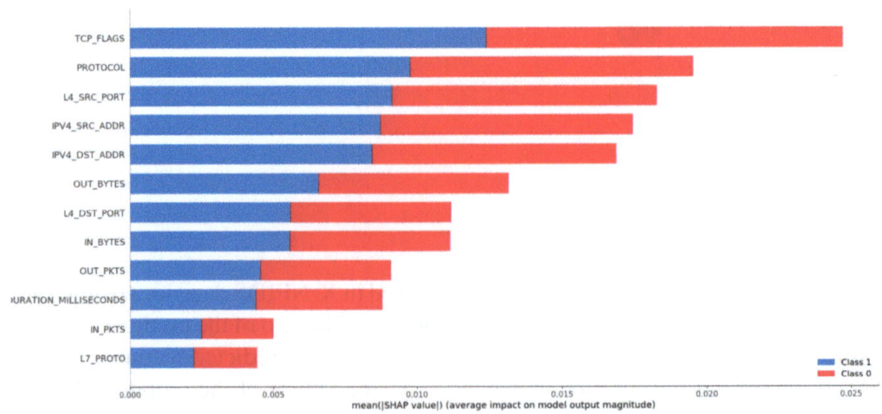

Fig. 5 The summary plot of X-NFIDS

5 Conclusion

In this chapter, we have presented an explainable NetFlow-based solution for intrusion detection in IoT networks. As NetFlow-based IDS depends on packet headers, it can reduce the volume of traffic and detect anomalies in encrypted traffic. The features extracted by NetFlow can help machine learning methods provide high intrusion detection performance. NetFlow-based IDS is a scalable method suitable for high-speed networks. Our proposed X-NFIDS was evaluated using two NetFlow datasets, and the results showed high accuracy. This chapter tested the feasibility of X-NFIDS in IoT networks. In our future work, different explanation models will be investigated to find an efficient combined model that can provide a more comprehensive explanation of IDS behaviour. In addition, the output results of the explanation models will be validated using multiple CPS datasets.

References

1. Akerman S, Habler E, Shabtai A (2019) VizADS-B: analyzing sequences of ADS-B images using explainable convolutional LSTM encoder-decoder to detect cyber attacks. https://doi.org/10.48550/ARXIV.1906.07921, https://arxiv.org/abs/1906.07921
2. Awad M, Fraihat S, Salameh K, Al Redhaei A (2022) Examining the suitability of NetFlow features in detecting IoT network intrusions. Sensors 22(16):6164
3. Breiman L (2001) Random forests. Mach Learn 45:5–32. https://doi.org/10.1023/A:1010933404324
4. Holzinger A, Saranti A, Molnar C, Biecek P, Samek W (2022) Explainable AI methods—a brief overview. In: International workshop on extending explainable AI beyond deep models and classifiers. Springer, pp 13–38
5. Jadidi Z, Muthukkumarasamy V, Sithirasenan E, Singh K (2016) A probabilistic sampling method for efficient flow-based analysis. J Commun Netw 18(5):818–825
6. Jadidi Z, Lu Y (2021) A threat hunting framework for industrial control systems. IEEE Access 9:164118–164130
7. Jadidi Z, Foo E, Hussain M, Fidge C (2022) Automated detection-in-depth in industrial control systems. Int J Adv Manuf Technol 118(7):2467–2479
8. Le TTH, Kim H, Kang H, Kim H (2022) Classification and explanation for intrusion detection system based on ensemble trees and SHAP method. Sensors 22(3):1154. https://doi.org/10.3390/s22031154, https://www.mdpi.com/1424-8220/22/3/1154
9. Lundberg SM, Erion G, Chen H, DeGrave A, Prutkin JM, Nair B, Katz R, Himmelfarb J, Bansal N, Lee SI (2020) From local explanations to global understanding with explainable AI for trees. Nat Mach Intell 2:56–67
10. Lundberg SM, Lee SI (2021) A unified approach to interpreting model predictions. In: Proceedings of the advances in neural information processing systems, pp 4765–4774
11. Mane S, Rao D (2021) Explaining network intrusion detection system using explainable AI framework. https://doi.org/10.48550/ARXIV.2103.07110, https://arxiv.org/abs/2103.07110
12. Marino DL, Wickramasinghe CS (2018) An adversarial approach for explainable AI in intrusion detection systems. https://ieeexplore.ieee.org/abstract/document/9555622
13. Millar J (2021) Principles and practice of explainable machine learning. Big Data. https://doi.org/10.3389/fdata.2021.688969
14. Muna RK, Maliha HT, Hasan M (2021) Demystifying machine learning models for IoT attack detection with explainable AI. http://dspace.bracu.ac.bd/xmlui/handle/10361/15553

15. Onose E (2021) Explainability and auditability in ML: definitions, techniques, and tools. Neptune.ai. https://neptune.ai/blog/explainability-auditability-ml-definitions-techniques-tools
16. Reyes AA, Vaca FD, Castro Aguayo GA, Niyaz Q, Devabhaktuni V (2020) A machine learning based two-stage Wi-Fi network intrusion detection system. Electronics 9(10). https://doi.org/10.3390/electronics9101689, https://www.mdpi.com/2079-9292/9/10/1689
17. Sarhan M, Layeghy S, Portmann M (2022) Towards a standard feature set for network intrusion detection system datasets. Mob Netw Appl 27(1):357–370
18. Sarhan M, Layeghy S, Moustafa N, Portmann M (2020) NetFlow datasets for machine learning-based network intrusion detection systems. In: Big data technologies and applications. Springer, pp 117–135
19. Sarhan M, Layeghy S, Portmann M (2021) Evaluating standard feature sets towards increased generalisability and explainability of ML-based network intrusion detection. https://doi.org/10.48550/ARXIV.2104.07183, https://arxiv.org/abs/2104.07183
20. Shapley LS (2016) A value for n-person games. Princeton University Press, p 17
21. Sinclair C, Pierce L, Matzner S (1999) An application of machine learning to network intrusion detection. In: Proceedings 15th annual computer security applications conference (ACSAC'99), pp 371–377. https://doi.org/10.1109/CSAC.1999.816048
22. Wali S, Khan I (2021) Explainable AI and random forest based reliable intrusion detection system
23. Wang M, Zheng K, Yang Y, Wang X (2020) An explainable machine learning framework for intrusion detection systems. IEEE Access 8:73127–73141. https://doi.org/10.1109/access.2020.2988359
24. Yang C, Chen M, Yuan Q (2021) The application of XGBoost and SHAP to examining the factors in freight truck-related crashes: an exploratory analysis. Accid Anal Prev
25. Zebin T, Rezvy S, Yuan L (2022) An explainable AI-based intrusion detection system for DNS over HTTPS (DoH) attacks. http://ray.yorksj.ac.uk/id/eprint/5892

Application of Machine Learning on Material Science and Problem Solving Under Security—A Review

Maedeh Beheshti and Jolon Faichney

Abstract In material science, understanding structure, property, performance, processing, characterization and those relationships is a main concern for experimental scientists. Also, discovering new materials to address some of crucial global challenges such as health and medicine, food and water security, climate, etc. raises other important issues for material scientists in terms of cost and time consuming. In this regard, combining data science and machine learning knowledge with the experts of material scientists can help us providing better guidance to solve these questions: What material to make? how to make them? And how to recognize their properties? This chapter is conducted into two separated parts applicable by machine learning models: (1) material science and (2) problem solving. First section of this chapter aims to introduce us with some progressing in this area. This section specifically considers automatic solutions for material science known as material informatics in order to prevent information security threats particularly information integrity threats that is probable in national security organizations. The second section attends to two useful graph-based models in problem solving. Graph-based models have been playing the most important role in unsolved problems such as computer vision, engineering, security and medicine for many years. Nowadays, there have been significant efforts in producing mature and improved graph-based algorithms. The aim of this research is to introduce two prevalent graph-based methods, namely, graph-cut models (deterministic) and a unified graphical model (probabilistic) in a simple word.

Keywords Material science · Machine learning · Chemistry · Graph-cut models · Probabilistic graphical methods · Max–min cut algorithms · Conditional random field

M. Beheshti (✉)
Critical Path Institute, Tucson, AZ 85718, USA
e-mail: maedehbeheshti79@gmail.com; mbeheshti@c-path.org

J. Faichney
School of Information and Communication Technology, Griffith University, Brisbane, Australia
e-mail: j.faichney@griffith.edu.au

S. Pal et al. (eds.), *Emerging Smart Technologies for Critical Infrastructure*,
Smart Sensors, Measurement and Instrumentation 44,
https://doi.org/10.1007/978-3-031-29845-5_5

1 Material Informatics Introduction

Material informatics is prime research that combines informatics and materials science to optimize the understanding of the usage, selection, development, and discovery of materials. As an emerging field, the goal is producing robust materials with reducing the time and risk required to develop. Nowadays, state-of-the-art machine learning tools allow us to classify and draw the decision boundaries more accurate than the use of pencil and paper approaches of last years.

Necessity of recognizing a function that allows us to predict the output value for a new set of input data accurately, encourages the scientists to be aware about the crucial role of training data, as well as any other prior knowledge. The training data which is a collection of known input (x) and output (y) values, may be generated through observations or controlled experiments. We introduced this research to provide a review knowledge of available strong infrastructure and framework of machine learning in material science in order to provide a high performance and security in our future products.

1.1 Machine Learning and Material Science

Nowadays machine learning is a significant paradigm to accelerate the discovery of material through an efficient and effective methods to generate, manage, and utilize relevant information [1, 2]. Mitigating cost, risks, and time involved in traditional, trial and error discovery of material is the main potential purpose of machine learning and artificial intelligence approaches for extraction beneficial knowledge of useful materials.

Figure 1 demonstrates an examples of machine learning application in material science.

Larger areas such as cognitive game theory (e.g., computer chess), pattern recognition (e.g., facial or fingerprint), event forecasting, and bioinformatics [2] are also other categories impacted by machine learning. They are considerably effective in materials science, materials research and discovery. Some kinds of recent examples of successful applications of machine learning in materials research are as below [2]:

- Accelerated and accurate predictions (using past historical data) of phase diagrams
- Crystal structures
- Materials properties
- Prediction of materials properties such as development of interatomic potentials
- Energy functionals for increasing the speed and accuracy of materials simulations
- On-the-fly data analysis of high-throughput experiments
- Mapping complex materials behavior to set of process variables
- Chemistry.

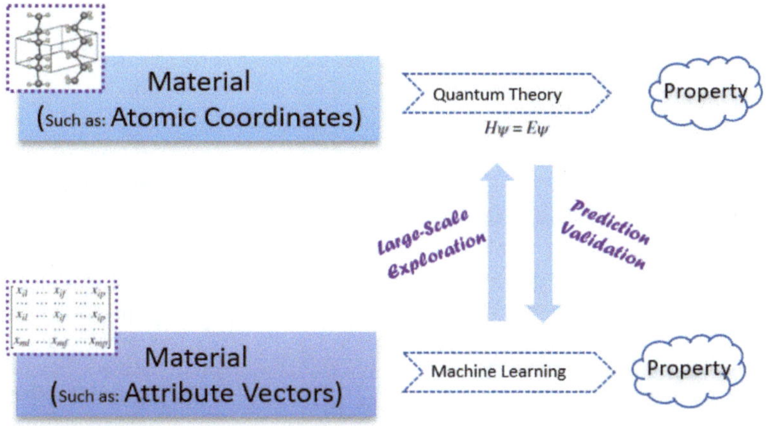

Fig. 1 Perspective of the role of machine learning on accelerating quantum mechanical computations

In [3] the authors proposed some mix design factors which influence the strength prediction of metakaolin-based geopolymer through experiments and statistical analyses.

Geopolymer is an important alternative binder to Portland cement. Although, there has been lots of endeavors in this area, this work mostly considers the gravity of the four common mix design parameters, namely Si/Al (molar ratio), water/solids (mass ratio), Al/Na (molar ratio) and H_2O/Na_2O (molar ratio), in regulating compressive strength of metakaolin-based geopolymers through experiments and statistical analyses. This paper represents the results of applying different types of machine learning algorithms such as Random forests, Naïve Bayes and k-nn on metakaolin-based geopolymer, with the purpose of classification. Attribute selection and evaluation along with strength prediction play serious role in assessing the significance of the mixed design parameters contributing to compressive strength.

Butler et al. [4] provide a summary of recent progress of machine learning in chemical sciences. Fundamental steps engaged in construction of a model, presented in Fig. 2. The authors considered a 4-step workflow procedure of computational chemistry.

1. Data Collection

Depending on the type of data, there are 3 types of supervised, semi-supervised and unsupervised machine learning. The most important subject that is a big concern in data collection is different kinds of error in data produced by human and any measurements. Lack of reproducibility along with error propagation of experimental data might happen as a result of mistakes in data [4, 5].

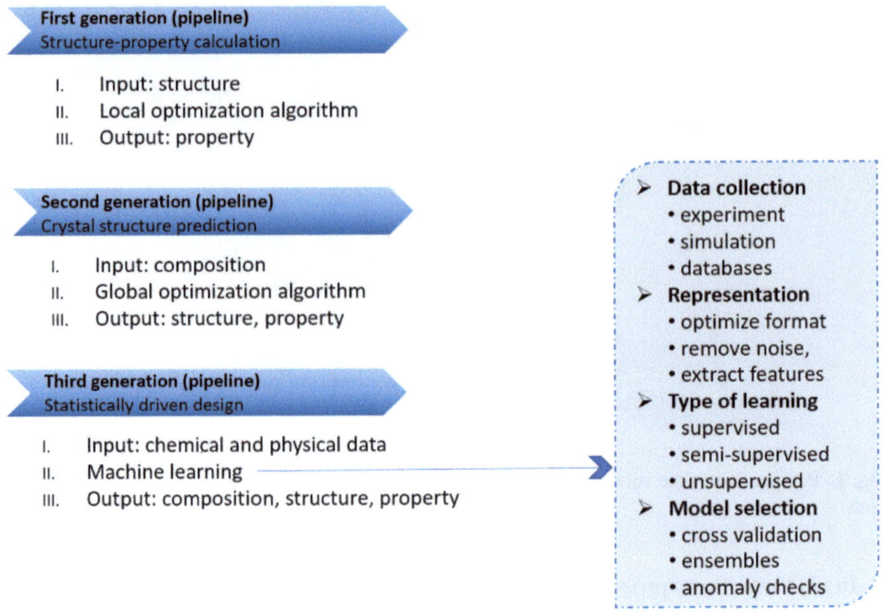

Fig. 2 Evolutional research workflow in computational chemistry

2. Data Representation

Representing raw data to an effective acceptable format for a machine learning algorithm is a data representation. The more suitable you define an input data representation, the more accurate you can expect to have the output of the machine learning. Data representation is still an open research problem for chemical systems [6–10].

3. Types of Learning

Selecting an appropriate learner is another phase after data representation. Figure 3 shows different models of supervised and unsupervised learning for chemical discovery based on type of data and question.

Two kinds of predictions for output results exist in supervised learning models:

(1) A discrete set that needs classification: as an example, categorizing a material into metal or insulator.
(2) A continuous set (such as polarizability) that requires regression.

Based on data type, even ensemble models can be proposed in some problems. Also, changing the internal parameters of the similar algorithms is a beneficial choice of creating a robust model.

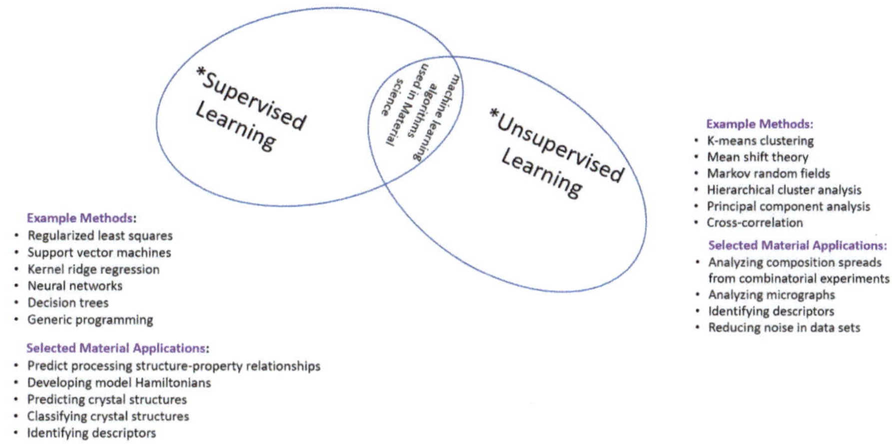

Fig. 3 An overview of different types of the machine learning algorithms in material science

4. Model Selection

After choosing the learner(s), a trial model for optimization and appropriate selection of the best model is needed to be evaluated. There are three main errors needed to be considered for choosing the best model that must be taken into account [4]:

(1) Model bias

This error results from incorrect hypothesis induced by the algorithm and can conclude with the model once losing substantial relationships. High bias or underfitting arises while the model:

(i) Cannot show adequate flexibility to explain the existing relationships of inputs and predicted outputs.
(ii) The data is not properly detailed to provide appropriate discovery of critical rules.

(2) Model variance

Instability and small variations in the training dataset can be identified as a variance. For machine learning models, even with well-trained procedures, variety of errors are inevitably possible. Those errors resulted from different factors like noise in the training data, measurement limitations, calculation uncertainties, simply outliers or missing data. High variance in terms of overfitting arises if a model tends to be too complex; specifically, at the time of increasing the number of parameters. A simple approach for recognizing the overfitting is the behavior of training and test datasets in terms of accuracy. In case of overfitting, the accuracy of a model in the training data is opposite of the test data. If one (training data) continues to show progress in accuracy, the other one (test data) estimate to decline in performance.

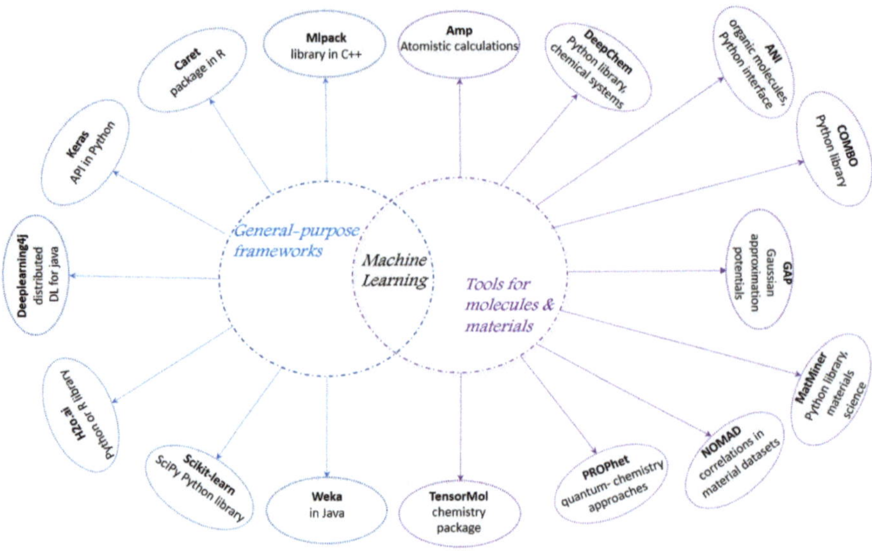

Fig. 4 Some available integrated material-machine learning tools

(3) Irreducible errors

Mathematical or statistical models help us to approximate reality for learning algorithms, that results in error calculation into two main categories:

(i) reducible error and (ii) irreducible error.

Reducible error: As the name represent, is able to be minimized further in order to maximize accuracy.

Figure 4 shows useful information of existing general machine learning applications which are suitable in material related programs.

Irreducible error (inherent uncertainty): Irreducible error and natural variability are combined together in a system. The irreducible error is possibly the result of measurement errors or inability to capture all the features of interest in addition to inherent variability in the data to model the underlying phenomenon in a better way. It means, it doesn't matter how well you provide your model, our data contains a certain amount of noise or irreducible error that is inevitable and difficult to be removed.

Meanwhile it worth to discuss two important concepts needed to provide better understanding of our models:

Underfitting: Models exhibiting small variance and high bias have a tendency to underfit the real target.

Overfitting: Models exhibiting high variance and low bias have a desire to overfit the real target.

Note: if your target truth is highly nonlinear, selecting a linear model to approximate it, will be concluded in introducing advancing a bias from the linear model's

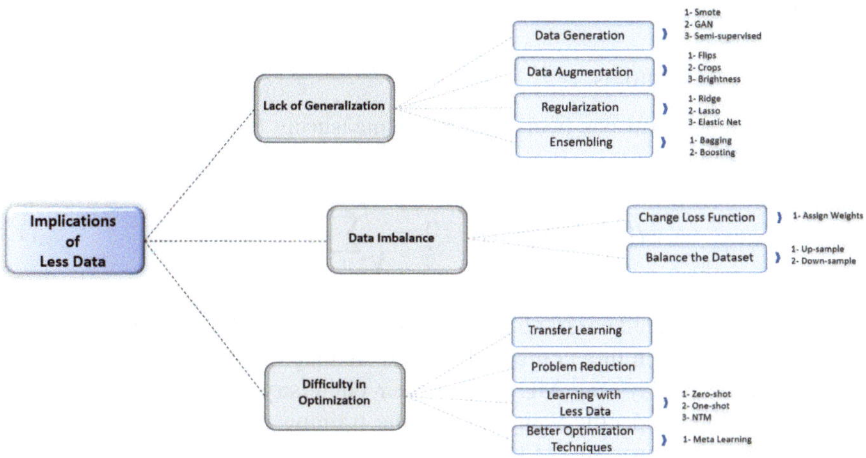

Fig. 5 Issues with data and analyses [11]

inability to apprehend nonlinearity. In other hand, your linear model is underfitting the nonlinear target function over the training set.

Figure 5 illustrates us about lack of data and consequences issues in analysis. The picture also provides us with some quick solutions in this area.

1.2 Application of Machine Learning in Chemistry

Chemical syntheses designed by practitioners tolerate several years of advanced training and experience in consequence of trial-and-error and labor-intensive fashion. As a result, the syntheses developed are often unreliable, difficult to scale, and frequently require revision in optimization or development. In this regard, a reformation is required in chemical synthesis design and development by proposing an integrated machine learning approach to the synthesis of any organic target molecule. The innovated auto-system (introduced as a combination of retrosynthetic tools, machine learning, and computational chemistry tools) allows users to focus on target selection instead of spending weeks, months, or years of effort now required for mapping a chemical idea into reality.

In [8] a machine learning model proposed for predicting the atomization energies of a diverse set of organic molecules. The proposed model is only relying on nuclear fluctuation and atomic positions. The aim is addressing the Schrodinger equation (SE), $H\Psi = E\Psi$, as an equational problem in mechanics for assemblies of atoms. Authors struggle with establishing a nonlinear map between molecular characteristics and atomization energies by using machine learning. In this regard, it is essential to define a relevant measurement of a molecular (dis)similarity, constant to below variations such as:

- translations,
- rotations,
- index ordering of atoms.

The Euclidean norm has been chosen for measuring the distance between two molecules as below:

$$d(M, M') = d(\epsilon, \epsilon') = \sqrt{\sum_I |\epsilon_I - \epsilon'_I|^2} \tag{1.1}$$

M is the diagonalized coulomb matrices of a molecule. The ML model proposed in [1, 12, 13] is chosen in a combined format. As they mentioned a support vector machine approach plays the ML role to their chemistry molecule-based models. In order to define a specific weight, α_i to the training molecule i, a kernel ridge regression, has been proposed [13].

The reason of the α_i is providing a parameter based on weight in addition to distance, for training molecule i in order to contribute to the energy.

1.3 Application of Machine Learning in Medicinal Chemistry

Over the past few decades, leveraging in silico QSAR (Quantitative structure–activity relationship) applications [14] with the purpose of predicting drug activities, attracted many experts in this area to apply them in drug discovery, repurposing and development [14]. As an example, two simple types of machine learning algorithms, multiple linear regression (MLR) and partial least squares (PLS) have been utilized on small data sets.

Binding assays is another interesting application in drug discovery, compelling experts to produce large amounts of data, considering the activity of diverse chemical matter. Random forests (RF), support vector machines (SVM), artificial neural network (ANN), and Cubist7 are only few examples of the emerging outstanding machine learning models in expanding different types of in silico ADMET and QSAR pipelines in prediction.

Hochuli et al. [15] proposed a convolutional neural network (CNN) to score protein binding. This score is important as a main part of a structure-based drug design procedures. Using CNN as a strong model of machine learning techniques in recognizing features and relationships helped them to provide an automatic model of choosing proper binding pose and affinity for protein–ligand complexes.

1.4 Image Processing and Material Science

As a branch of machine learning, image processing has a breakthrough innovation in the material science. Material clustering, recognition, content-based construction and automation are the most important application of computer vision in the material science area. Gácsi [16] proposed an image processing approach for recognizing 3-D structure of materials. In this model authors extract the properties related to material based on a mathematical approach named stereometric microscopy. By projecting a 3D model to 2D, three dimensional properties of the material will be investigated in a two-dimensional image. In other words, a quantitative interpretation of image information and content, in terms of image analysis, is fulfilled.

Figure 6 represents six images of grain that can be considered in image processing for classification.

Fig. 6 ISO-945 defined six classes of grains in cast iron [17]

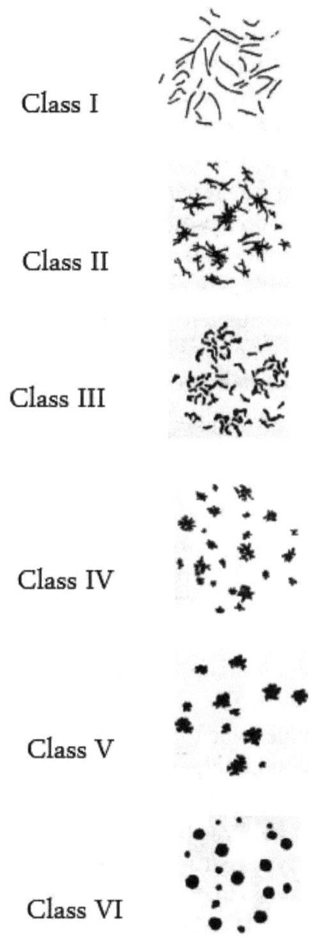

1.4.1 Exploring the MAX Ternary for Extracting Elastic, and Electronic Properties

In [18, 19] a data mining approach proposed to investigate the stability, elastic, and electronic characteristics of the MAX phases. MAX phases or Mn + 1 A Xn are the anisotropic laminated transition metal composed with hexagonal crystal structure. They are a class of ternary compounds containing an unusual composition of mechanical and electronic properties with high variety that allows investigating for the strong elasticity and electronic structure.

The focus of the Aryal et al. work is on the elastic features, electronic structure, correlations and integrities between the two, for Mn + 1 A Xn phases (layered hexagonal crystals) with n varying from 1 to 4.

Depending on M, A, X, and n, the elastic properties of MAX phases will be different in a broad range. Two critical mechanical properties of the MAX phases are the Poisson's ratio ŋ, which is closely related to Pugh moduli ratio (G/K), and the total bond order density (TBOD), which is an indicator of the strength of interatomic bonding in a crystal. G/K ratio (the Poisson's ratio or the Pugh moduli ratio) represents a balanced overall assessment of the mechanical properties and is an indication of a possible increase in its relative ductility in pure metals. Bulk (G) is resistance to change in volume and shear moduli (K) is resistance to change in shape, are the most important parameters.

A data set including 665 MAX phase properties used to test the influence of data-mining statistical learning approach to materials property predictions (G, K G/k). A multiple linear regression of the simplest data-mining method implemented in Weka environment proposed for this coding.

One of the major issues in applying AI to materials science is a lack of data, especially compared to other application areas of AI, where vast amounts of data are already available or can be obtained cheaply. As a result, with no doubt, a large part of the relevant literature is not much concerned with the urgency of the application of AI techniques. As we mentioned the difficulty with data shortage in this area is important issue, and encourages the researchers to mostly focus on creating the required preconditions necessary for providing datasets on materials and their properties, that are expensive to compile, but mandatory for the application of more advanced AI techniques. References [18, 19] consider this in the context of MAX phase materials. These materials are of particular interest to the materials science community because of the wide range of different properties they have depending on the exact composition. For example, they can behave like metals or ceramic, able to be tolerant to damage, resistant to oxidation, while electrically and thermally conductive. MAX phase materials are of interest for example in applications that expose them to extreme conditions.

MAX phase materials comprise layered carbides and nitrites with a hexagonal crystal structure. The exact elements for the M, A, and X layers, as well as the layer structure, are varied for different materials and give rise to different properties. The theoretical relationship between the structure and composition of a MAX phase material with its properties is currently poorly understood. That's a reason of preparing the area appropriate for the application of machine learning techniques that can help us to provide predictions with better understanding which cannot currently be achieved based on our available understanding of the physics.

The authors applied linear regression to this dataset of 665 materials. The properties of the electronic structure used as features to predict the mechanical properties of the same material. In particular, the machine learning relates the bulk modulus (the ratio of the change in pressure to the fractional volume compression) and Poisson's ratio (the ratio of transverse strain to axial strain, a measure of how much a material expands perpendicular to compression) to the electronic properties.

2 A Review on Two Graph-Based Models for Problem Solving in a Simple Language

Graph-based algorithms can be largely partitioned into two types (1) deterministic and (2) probabilistic. Algorithms such as max/min cut and normalized cut constitute the deterministic approach, and algorithms such as Conditional Random Field (CRF) and Bayesian Network establish the probabilistic graphical approach. Energy minimization and inference provide the main parts in graph cut algorithms and probabilistic models, respectively. There are advantages and disadvantages for each graph-based algorithm. For example, the necessity of dealing with many variables in energy minimization approaches, resulted in a highly complex energy function of graph-cut algorithms and often NP-hard, especially for non-convex problem. However, there are different applications for deterministic and probabilistic methods based on the types of applications and requirements. As an example, traffic engineering may receive benefits from max/min algorithms to identify the maximum flow rate of vehicles from the downtown car park to the freeway. This will have an impact on traffic engineers' decisions to widen the roadways. Another example can be the medical image segmentation with maximum accuracy to help doctors for making the best decisions about their patient's disease situations.

Sections 2.1 and 2.2 illustrate the performance of the two approaches (graph-cut and probabilistic models) in a simple use case for images with complex background scenes.

2.1 How Graph-Cut Models Work for Image Segmentation?

Maximum flow algorithms offer principal part of many global optimization methods like graph-cut models. There are many methods for implementing the maximum flow algorithms, such as Goldberg–Tarjan's [20] "push-relabel" methods and algorithms based on Ford–Fulkerson [21]. In this section, we will describe the Ford–Fulkerson algorithm due to its effectiveness and popularity in many graph-cut methods.

Figures 8, 9, 10 and 11 show the maximum flow/minimum cut procedure for finding the maximum flow from S (source) node to T (terminal) node [20]. The weights shown in the initial weighted network, Fig. 8a, indicate the capacity of the edges, and the current maximum flow from one node to the other node. Figure 8b represents the equal residual network for the capacity network. For example, in Fig. 8a the maximum flow from node (2) to node (4) is 4. There are many paths from S to T in Fig. 8a and each path has a bottleneck edge. The edge in a path which has a minimum capacity is a bottle neck edge. The purpose of max/min cut is to determine the maximum amount of flow which can be sent from one source node S to the terminal node T. For example, in Fig. 8a although the capacity of the (S1) edge is 7, the maximum flow of 5 can be sent from S to T in a (S1T) path. Therefore, the (1T) edge is a bottle neck here. Each iteration tries to saturate at least one path in the graph. This means that after each saturation, all the capacity of the edges constructed for that path has been used. Therefore, the maximum amount of flow is expected to be passed through a saturated path. In each iteration, we try to find a path from S to T. If there is an edge in the original graph with capacity of 3 as depicted in Fig. 7a, its equal edge in the residual network is according to Fig. 7b. For each edge there is a flow/capacity number such as 3/0 in residual graph. Algorithm 2.1 shows how capacity in the residual network is calculated.

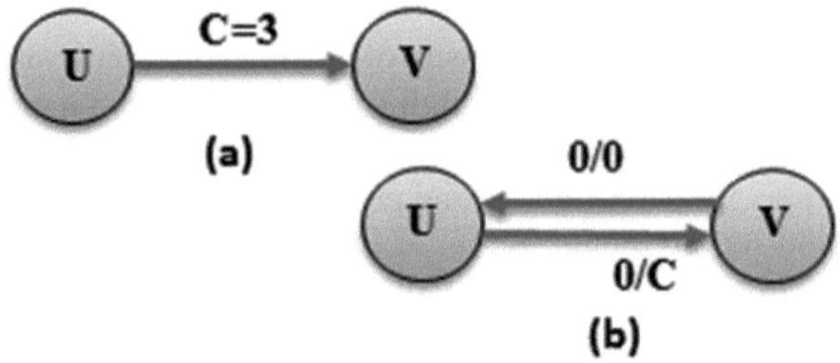

Fig. 7 **a** Edge in the original graph and **b** edge in the residual network [20]

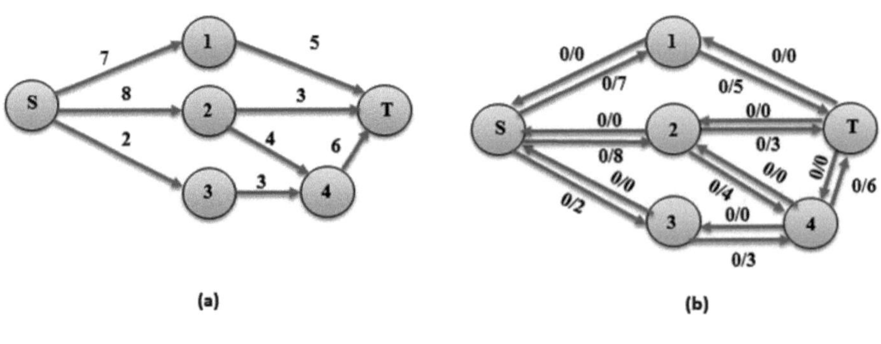

Fig. 8 **a** Initial network and **b** residual network [20]

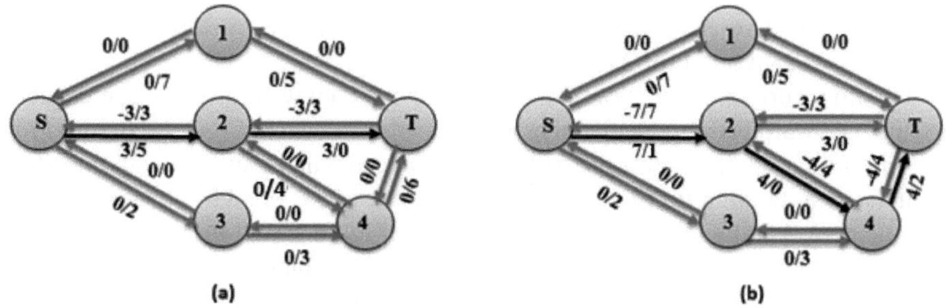

Fig. 9 **a** Iteration 1 (S2T) and **b** iteration 2 (S24T) [20]

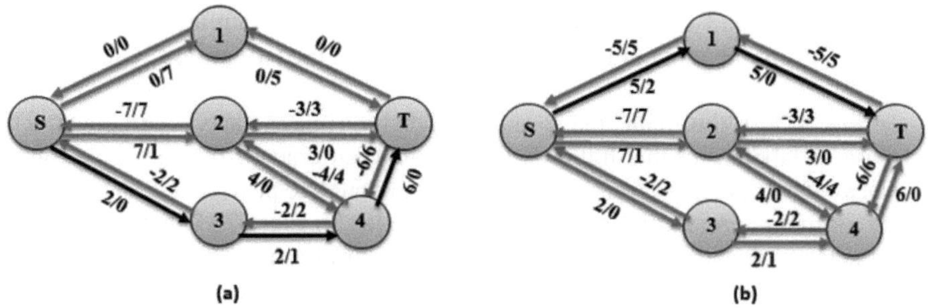

Fig. 10 **a** Iteration 3 (S34T) and **b** iteration 4 (S1T) [20]

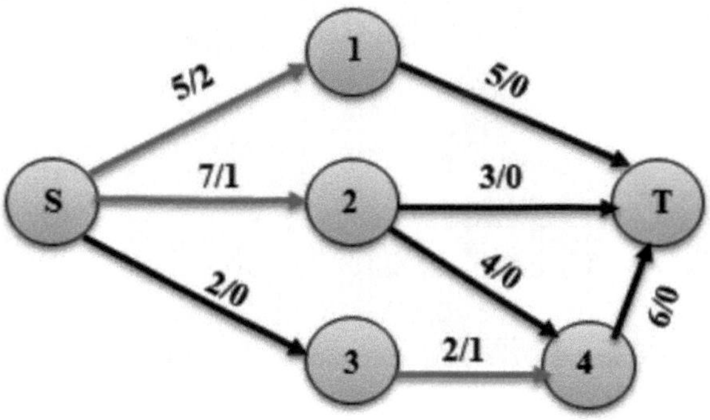

Fig. 11 Final flow graph [20]

In this use case, beginning from iteration 1, which has been shown in Fig. 8, all the procedure will be finished in iteration 4. In each iteration, a path will be saturated. Path S2T has been selected for the first time in Fig. 8. In this path the capacity between 2 and T nodes, which is 3 units, shows a bottle neck.

Figure 9a shows how the capacity for residual network is calculated according to Algorithm 2.1. For the forward arrows in the S2T path in the residual network, (0 +

3) and (8 − 3) show 3/5 calculation for the S2 forward arrow. Also (0 + 3) and (3 − 3) show 3/0 calculation for 2T forward arrows.

The flow/current calculation is the same for the backward arrows. The capacity for the backward path is calculated as *Capacity After Flow* = *Capacity Before Flow* − *Flow*. Figure 9a represents the path found between S and T as S2T with max flow, 3. In the next iteration for finding the saturated path S24T, Fig. 9b shows a path with a bottle neck 4 and a maximum achieved flow 4. Figure 10a, b show S34T and S1T saturations which resulted in 2 and 5 maximum flows. Finally, the maximum flow is calculated as 3 + 4 + 2 + 5 = 14. As Fig. 11 depicts, there are two ways from S to reach one of the nodes with number of 1 or 2, but there is no way for them to reach T node. All the nodes to T node have been blocked and their capacity has been used completely. At this point, we say that all the paths from S to T have been saturated. In this regard, S and T will be separated by removing some saturated edges, 1T, 2T, 24 and S3 from the main graph. The separated nodes are two groups, $\{S, 1, 2\}$ and $\{3, 4, T\}$.

In Fig. 12a, b the main image and the labeling image have been shown. The purpose of labeling in this figure is to determine an object from the background. Then two labels (same as S and T terminals) are enough to provide this classification based on max–min algorithms of graph-cut model. In Fig. 12b it has been shown how the main object received a different label (white) from the background (black).

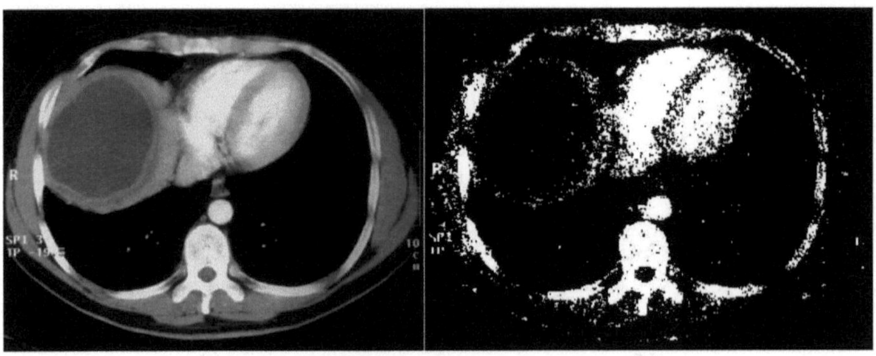
(a) (b)

Fig. 12 **a** The abdomen image, **b** the labeled image of abdomen [22]

Algorithm 2.1 Ford–Fulkerson

```
Ford-Fulkerson Algorithm G (V, E, s, t, fw, ca)
fw(u,v)=0 for each pair (u,v) in V*V              // Initialization term
Construct residual network GR
  Residual network GR(V,ER)= G(V,E)
   ∀ (u,v) ∈ E ,  (u,v)∈ ER and (v,u) ∈ ER  // Initialization term
     fwR(u,v)= 0 and fwR(v,u)=0              // Initialization term
    caR(u,v)= ca(u,v) - fwR(u,v)            // Initialization term
     caR(v,u)= ca(v,u) - fwR(v,u)          // Initialization term
While there is an s-t path p in the residual network GR
 & caR(u,v)>0 for all edges(u,v)
      do  // increase flow by augmenting flow along p
               caR (p) ←——  min caR(u,v)
              fwR(u,v) ←—— fwR(u,v)+caR(p)
              caR(u,v) ←—— caR(u,v) - fwR(u,v)
              fwR(v,u) ←—— fwR(v,u)- caR(p)
              caR(v,u) ←—— caR(v,u) - fwR(v,u)
                fw     ←—— fw +caR(p)
           return f
  End.
```

2.2 *How Conditional Random Field Address Image Segmentation?*

In this section, an example of a simple undirected probabilistic graphical model with the concept of CRF is introduced [23–25] to predict the percentage of cheating in a

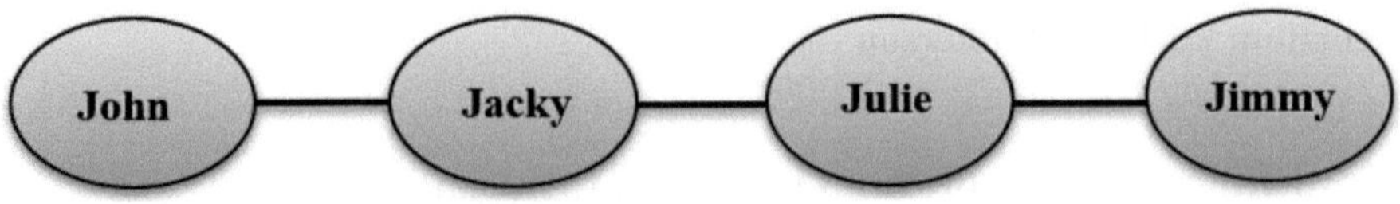

Fig. 13 Order of seating for four students [26]

four students' room. John, Jacky, Julie and Jimmy, are all the students seated in one room in the order shown in Fig. 13.

All students can see the answers of the others and everybody thinks that the neighboring student studied more, and therefore has the correct answers. However, only Jacky and Jimmy studied well for their exam. Due to this presumption that all studied well, all of their answers are closely related to each other. Although John and Julie are far from each other, their answers are dependent due to Jacky. In an undirected graphical model (UGM) two variables are dependent on each other if there is a path between them. So, all of the four variables in the example are dependent on each other. But there is an exception here in the concept of conditionally independent variables. For example, even though Jimmy and Jacky are dependent, they will be independent if we know the value of Julie. Now a pairwise UGM model from this example concerning dependencies and independencies is constructed. According to Eq. 2.1, the joint probability of a particular assignment to all of the variables x_n will be shown as a normalized product of a collection of non-negative factors (function).

$$p(x1, x2, \dots, xN) = \frac{1}{Z}\prod_{i=1}^{N} \psi_n(x_n)\prod_{e=1}^{E} \psi_e(x_{ek}, x_{el}) \tag{2.1}$$

ψ_{n} is a factor or function for node n, and ψ_e is a factor for edge e. In this example, the nodes are students and the edges are the only link between two students who are seated next to each other. There is no edge between variables with one or more nodes distance. ψ_n is a node potential function which gives a non-negative weight to each possible value of the random variable x_n.

For example, $\psi_3(W\_ans) = 65$, and $\psi_3(C\_ans) = 35$ means that the third node's (Julie) answers tend to be more incorrect than correct. Similarly, ψ_e is an edge potential which gives a non-negative weight to each possible value of the random variable of edges. Edge is a combination of two node variables x_{ek} and x_{el}. For example, the students who are seated next to each other receive a higher edge score than those who are far from each other. Z is a normalization constant which gives the above distribution a summation of one over all possible joint configurations of the variables. The definition of Z is as Eq. 2.2.

$$Z = \sum_{x_1}\sum_{x_2}\cdots\sum_{x_n}\prod_{n=1}^{N} \psi_n(x_n)\prod_{e=1}^{E} \psi_e(x_{ek}, x_{el}) \tag{2.2}$$

Table 1 Node potentials

Student	Correct answer	Wrong answer
John	0.25	0.75
Jacky	0.9	0.1
Julie	0.25	0.75
Jimmy	0.9	0.1

Table 2 Edge potentials

n1/n2	Correct answer	Wrong answer
Correct answer	2/6	1/6
Wrong answer	1/6	2/6

In this example of UGM, there are 4 nodes corresponding to 4 students and each node has 2 states, correct and incorrect. Table 1 shows how node potentials have been assigned to each node.

Table 2 shows how edge potentials have been assigned to each edge. The potential amount for having the same state for any two neighbor nodes is two times more than having different state.

Table 3 shows probability distribution by UGM aligned with all possible configurations of Correct/Incorrect (wrong) answers for each student. N_p and E_p are summarizations of node potential and edge potential respectively. According to Eq. 2.1 the Pro-cul column is calculated by multiplying all node and edge potentials in a row.

Also, the last column, Probability distribution, is the result of (Pro-cul)/Z. Z, as appears in Eq. 2.2 is obtained by the summation of the Pro-cul column. Noted that in order to simplifying the probability calculation in Table 3, all the numbers in each row in Tables 1 and 2 have been multiplied with a number, such as 4 or 6 to give an integer, and this number of multiplications for each row can be different.

Figure 14 represents a different condition in which all students were seated in different rooms. It shows the real score which each student can achieve without cheating. Each column shows a student and each row depicts one question. Blue color represents a correct answer and red color is the wrong answer. For example, as shown in Fig. 14, Jacky and Jimmy, who studied well, achieved the best results for 200 questions. In contrast, John and Julie who did not study well did not achieve good results.

Figure 15 shows a sampling of 200 questions from the configurations in Table 3. As depicted in this figure, John and Julie achieved extra correct answers by cheating, and also Jacky and Jimmy lost more correct answers because of cheating. In order to achieve the sampling of Fig. 15, one random number between [0, 1] was chosen.

For example, suppose 0.65 was selected. Referring to the Table 3, 0.17 was achieved for the first configuration from the probability column; then $0.17 + 0.26 = 0.43$ for the second configuration; $0.17 + 0.26 + 0 = 0.43$ for the third configuration; $0.17 + 0.26 + 0 + 0.03 = 0.46$ for the fourth configuration; $0.17 + 0.26 + 0 + 0.03 + 0.13 = 0.59$ for the fifth configuration; and finally, $0.17 + 0.26 + 0 + 0.03$

Table 3 Probability of each configuration for four students

John	Jacky	Julie	Jimmy	Np(1)	Np(2)	Np(3)	Np(4)	Ep(1)	Ep(2)	Ep(3)	Pro-cul	Probability
C_ans	C_ans	C_ans	C_ans	1	9	1	9	2	2	2	648	0.17
W_ans	C_ans	C_ans	C_ans	3	9	1	9	1	2	2	972	0.26
C_ans	W_ans	C_ans	C_ans	1	1	1	9	1	1	2	18	0.00
W_ans	W_ans	C_ans	C_ans	3	1	1	9	2	1	2	108	0.03
C_ans	C_ans	W_ans	C_ans	1	9	3	9	2	1	1	486	0.13
W_ans	C_ans	W_ans	C_ans	3	9	3	9	1	1	1	729	0.19
C_ans	W_ans	W_ans	C_ans	1	1	3	9	1	2	1	54	0.01
W_ans	W_ans	W_ans	C_ans	3	1	3	9	2	2	1	324	0.09
C_ans	C_ans	C_ans	W_ans	1	9	1	1	2	2	2	36	0.01
W_ans	C_ans	C_ans	W_ans	3	9	1	1	1	2	2	54	0.01
C_ans	W_ans	C_ans	W_ans	1	1	1	1	1	1	2	1	0.00
W_ans	W_ans	C_ans	W_ans	3	1	1	1	2	1	2	6	0.00
C_ans	C_ans	W_ans	W_ans	1	9	3	1	2	1	1	108	0.03
W_ans	C_ans	W_ans	W_ans	3	9	3	1	1	1	1	162	0.04
C_ans	W_ans	W_ans	W_ans	1	1	3	1	1	2	1	12	0.00
W_ans	W_ans	W_ans	W_ans	3	1	3	1	2	2	1	72	0.02

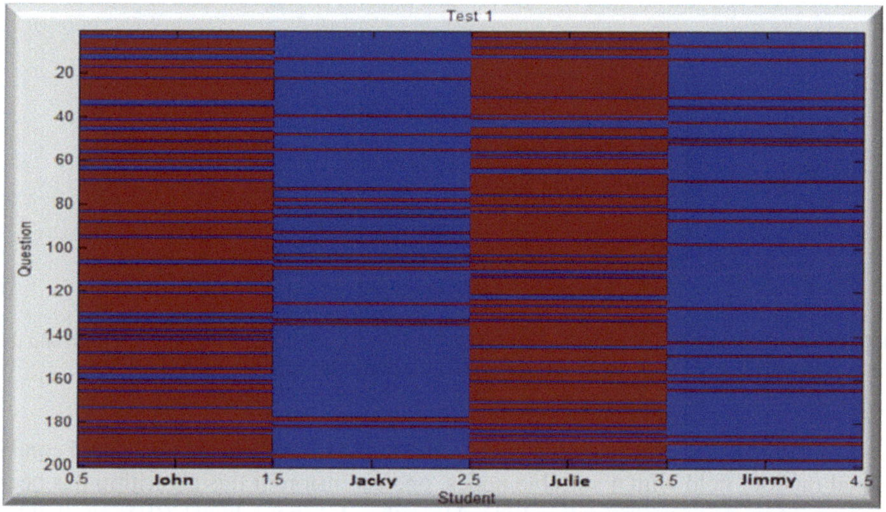

Fig. 14 A sample of cheating for 200 questions for four students seated in different rooms [23, 27]

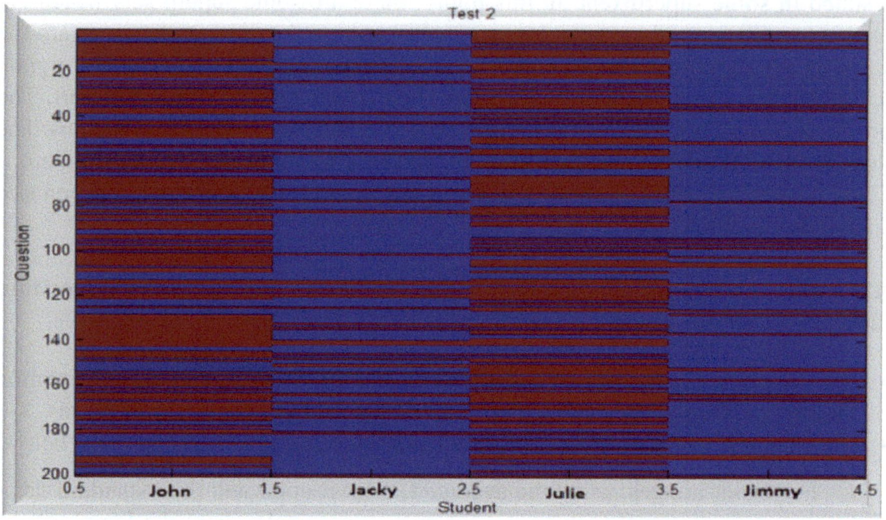

Fig. 15 A sample of cheating for 200 questions for four students seated in the same room [23, 27]

$+ 0.13 + 0.19 = 0.78$ for the sixth configuration. As 0.78 exceeded the threshold of 0.65, we stopped the calculation. The achieved configuration, shown in the sixth row of Table 3 is, w_ans, c_ans, w_ans, c_ans. If this process for the calculations is carried out for additional configurations up to 200 questions, the result will be shown in Fig. 15.

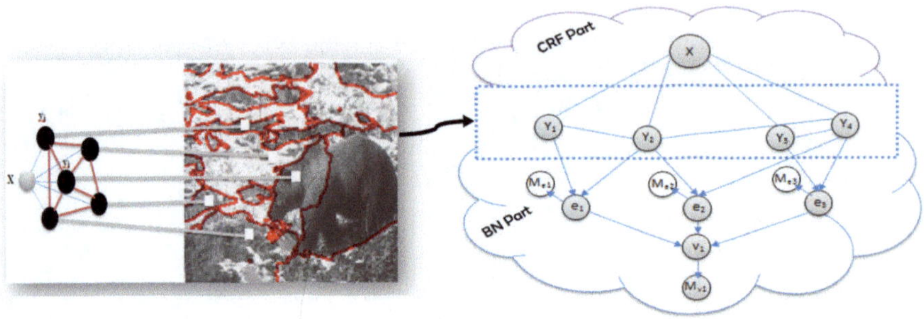

Fig. 16 An over-segmented image with related super-pixels which show the nodes in CRF and the links between them are the connections between the nodes [29, 30]

Regarding the above example which introduced us simply to the concept of conditional random field with 4 variable nodes, one extension of this algorithm has been proposed for image segmentation [28]. Algorithm 2.2 shows how CRF is applied for image segmentation. The concept behind the algorithm is exactly the same as the example of the students above. Figure 16 shows how an oversegmentation method resulted in some superpixels. In this figure, there are some superpixels which are mapped to CRF variables, and the extracted features like color and texture are the value of any superpixel. Each superpixel (specified with white color) in the original image has been mapped with one region node (black color) in the CRF part. The grey node x represents the whole image and y_i, y_j represent the labels of regions. The links in the CRF model obtained from the pairwise relationship. It means if there is a link between two adjacent superpixels, a link also will be appeared between two CRF nodes in the model.

Algorithm 2.2 Image Segmentation with a Unified Graphical Model Algorithm [28]

- Over-segment the image and extract super-pixels using Watershed Segmentation Method or Normalized-cut method (No. of super-pixels can be 15 or more).
- Node measurement using feature vector extracted from each superpixel as below:

  1. For Color images: using the average CIELAB color and their standard deviations as the local features x_i (for each superpixel)—here the length of the feature vector is 6
  2. Because each node has to be labeled, a 3-layer perceptron with an structure of 6 nodes in the input layer, 35 nodes in the hidden layer and 1 node in the output layer is used
  3. For Gray-Scale images: using the average intensity and 12 Gabor textures as the local features x_i (for each superpixel)—here the length of the feature vector is 13—the Gabor textures are calculated by filtering the gray-scale image with a set of Gabor filter banks. The average magnitude of the filtered image in

each superpixel region is used as the Gabor feature. Gabor filter banks with 3 scales and 4 orientations are used
4. Because each node has to be labeled, a 3-layer perceptron with a structure of 13 nodes in the input layer, 25 nodes in the hidden layer and 1 node in the output layer is used
5. Output of classifier is + 1 or − 1. It is a classifier based on a multilayer perceptron (MLP)
6. CRF Model: $\frac{1}{Z}\prod_{i\in V}\phi(y_i, x_i)\prod_{i\in V}\prod_{j\in N_i} exp(y_i y_j \lambda^T g_{ij}(x))$, as illustrated in the above example, the first part is the node potential and second part is the edge potential
7. $\phi(y_i, x_i) = \frac{1}{1+exp\left(-y_i\frac{net(x_i)}{\tau}\right)}$ ($Net(x_i)$ is calculated by Neural Network).

$g_{ij}(x) = [1, |x_i - x_j|]^T$, x_i and x_j are feature vectors of node i and j respectively. Maximum likelihood estimation (MLE) is used to find parameter λ.

3 Conclusion

In conclusion, due to the important role of statistics and machine learning in reproducing effective and beneficial materials in national security, the procedure of integrating material knowledge with informatics in terms of improving productivity and robustness is a critical subject. In our multipolar world, materials mater for each part of nations. Their usage is growing in aerospace, industrial, military, medicine and etc. so national security concerns have been changed thoroughly to this area in recent years. In summary, the second part of this chapter, helps us understanding the most important graph base algorithms in a simple word. We can observe the role of those algorithms in some critical applications such as flight scheduling in airlines, medical image segmentation, gaming and etc. that need high security in structure and backbone.

References

1. Müller K-R, Mika S, Rätsch G, Tsuda K, Schölkopf B (2001) An introduction to kernel-based learning algorithms. IEEE Trans Neural Netw 12(2):21
2. Wei J et al (2019) Machine learning in materials science. InfoMat 1(3):338–358. https://doi.org/10.1002/inf2.12028
3. Lahoti M, Narang P, Tan KH, Yang E-H (2017) Mix design factors and strength prediction of metakaolin-based geopolymer. Ceram Int 43(14):11433–11441. https://doi.org/10.1016/j.ceramint.2017.06.006
4. Butler KT, Davies DW, Cartwright H, Isayev O, Walsh A (2018) Machine learning for molecular and materials science. Nature 559(7715):547–555. https://doi.org/10.1038/s41586-018-0337-2
5. Fourches D, Muratov E, Tropsha A (2010) Trust, but verify: on the importance of chemical structure curation in cheminformatics and QSAR modeling research. J Chem Inf Model 50(7):1189–1204. https://doi.org/10.1021/ci100176x

6. Faber FA et al. Supplemental materials for 'Prediction errors of molecular machine learning models lower than hybrid DFT error', p 17
7. Isayev O, Oses C, Toher C, Gossett E, Curtarolo S, Tropsha A (2017) Universal fragment descriptors for predicting properties of inorganic crystals. Nat Commun 8(1):15679. https://doi.org/10.1038/ncomms15679
8. Rupp M, Tkatchenko A, Müller K-R, von Lilienfeld OA (2012) Fast and accurate modeling of molecular atomization energies with machine learning. Phys Rev Lett 108(5):058301. https://doi.org/10.1103/PhysRevLett.108.058301
9. Ward L et al (2017) Including crystal structure attributes in machine learning models of formation energies via Voronoi tessellations. Phys Rev B 96(2):024104. https://doi.org/10.1103/PhysRevB.96.024104
10. Schütt KT, Glawe H, Brockherde F, Sanna A, Müller KR, Gross EKU (2014) How to represent crystal structures for machine learning: towards fast prediction of electronic properties. Phys Rev B 89(20):205118. https://doi.org/10.1103/PhysRevB.89.205118
11. Brown AW, Kaiser KA, Allison DB (2018) Issues with data and analyses: errors, underlying themes, and potential solutions. Proc Natl Acad Sci USA 115(11):2563–2570. https://doi.org/10.1073/pnas.1708279115
12. Scholkopf B (2006) Max-Planck-Institut für biologische Kybernetik, Tübingen, Germany, p 179
13. Tibshirani S, Friedman H. Valerie and Patrick Hastie, p 764
14. Tsou LK et al (2020) Comparative study between deep learning and QSAR classifications for TNBC inhibitors and novel GPCR agonist discovery. Sci Rep 10(1):16771. https://doi.org/10.1038/s41598-020-73681-1
15. Hochuli J, Helbling A, Skaist T, Ragoza M, Koes DR (2018) Visualizing convolutional neural network protein-ligand scoring. J Mol Graph Model 84:96–108. https://doi.org/10.1016/j.jmgm.2018.06.005
16. Gácsi Z (2003) The application of digital image processing to materials science. MSF 414–415:213–220. https://doi.org/10.4028/www.scientific.net/MSF.414-415.213
17. Prakash P, Mytri VD, Hiremath PS (2011) Comparative analysis of spectral and spatial features for classification of graphite grains in cast iron. Int J Adv Sci Technol 29:10
18. Ching Y, Aryal S, Sakidja R, Barsoum MW. A genomic approach to study the properties and correlations of MAX phases, p 34
19. Aryal S, Sakidja R, Ouyang L, Ching W-Y (2015) Elastic and electronic properties of $Ti_2Al(C_xN_{1-x})$ solid solutions. J Eur Ceram Soc 23
20. Goldberg AV. A new approach to the maximum-flow problem, p 20
21. Delong A, Boykov Y (2008) A scalable graph-cut algorithm for N-D grids. In: 2008 IEEE conference on computer vision and pattern recognition, Anchorage, AK, June 2008, pp 1–8. https://doi.org/10.1109/CVPR.2008.4587464
22. Luo Q et al (2013) Segmentation of abdomen MR images using kernel graph cuts with shape priors. BioMed Eng OnLine 12(1):124. https://doi.org/10.1186/1475-925X-12-124
23. Koller D, Friedman N (2009) Probabilistic graphical models: principles and techniques. MIT Press, Cambridge, MA
24. Schmidt M, Murphy K, Fung G, Rosales R (2008) Structure learning in random fields for heart motion abnormality detection. In: 2008 IEEE conference on computer vision and pattern recognition, Anchorage, AK, June 2008, pp 1–8. https://doi.org/10.1109/CVPR.2008.4587367
25. Schmidt M, Alahari K (2011) Generalized fast approximate energy minimization via graph cuts: alpha-expansion beta-shrink moves. arXiv:1108.5710 [cs], Aug 2011. Accessed 01 May 2022. [Online]. Available: http://arxiv.org/abs/1108.5710
26. Sun Z, Li Z, Wang H, Lin Z, He D, Deng Z-H (2020) Fast structured decoding for sequence models. arXiv, 09 Jan 2020. Accessed 10 Oct 2022. [Online]. Available: http://arxiv.org/abs/1910.11555
27. Bishop CM (2006) Pattern recognition and machine learning. Springer, New York
28. Zhang L, Ji Q (2010) Image segmentation with a unified graphical model. IEEE Trans Pattern Anal Mach Intell 32(8):1406–1425. https://doi.org/10.1109/TPAMI.2009.145

29. Beheshti M, Liew AW-C (2014) Image segmentation based on graph-cut models and probabilistic graphical models: a comparative study. In: Wang X, Pedrycz W, Chan P, He Q (eds) Machine learning and cybernetics, vol 481. Springer Berlin Heidelberg, Berlin, Heidelberg, pp 371–378. https://doi.org/10.1007/978-3-662-45652-1_37
30. Beheshti M, Ashapure A, Rahnemoonfar M, Faichney J (2018) Fluorescence microscopy image segmentation based on graph and fuzzy methods: a comparison with ensemble method. IFS 34(4):2563–2578. https://doi.org/10.3233/JIFS-17466

Introduction to Blockchain Technology with Bitcoin Protocol

Babu Pillai, Jeyakumar Samantha Tharani, Zhé Hóu, Kamanashis Biswas, and Vallipuram Muthukkumarasamy

Abstract The engineering behind the technology that powers Bitcoin, known as Blockchain, has gained attention as a potential software solution for various industrial applications. The capability of revolutionising digital transactions brought significant interest in this technology and evolved greatly in the past decade. However, the development of applications beyond cryptocurrency has not yet kept pace. In this chapter, we will examine the basic design principles of blockchain technology using Bitcoin's architecture as a foundation and understand the rationale behind its design and the limitation of scalability.

Keywords Distributed ledger technology · Blockchain · Blockchain scalability trilemma

1 Overview of Distributed Ledger Technology

Distributed Ledger Technology (DLT) is a rapidly developing concept that deals with the process and technologies enabling a network of nodes to reach an agreement on the status of a distributed ledger [1]. Unlike traditional databases, there's no central storage or administration in a distributed ledger; instead, it has a set of pseudo-synchronized databases among a network of nodes. DLT operates based on distributed consensus, where each node can verify, process, and confirm every transaction [2]. The transactions are then recorded in the distributed ledger and synchronized across the network once a consensus is reached. The challenge lies in ensuring that all nodes maintain the same record in their ledgers, despite potentially *selfish*, *faulty*, or *malicious* nodes in the network.

B. Pillai (✉)
Southern Cross University, Bilinga, Australia
e-mail: pillai.babu@outlook.com

B. Pillai · J. S. Tharani · Z. Hóu · V. Muthukkumarasamy
Griffith University, Brisbane, Australia

K. Biswas
Australian Catholic University, Brisbane, Australia

S. Pal et al. (eds.), *Emerging Smart Technologies for Critical Infrastructure*, Smart Sensors, Measurement and Instrumentation 44,
https://doi.org/10.1007/978-3-031-29845-5_6

Definition 1 (*Distributed system*) The distributed system is a collection of independent entities that communicate through a communication medium to achieve a common goal and appear as a single coherent system to its users.

The entities in DLT are referred to as nodes. They can either be hardware devices or software processes. In practice, they are independent nodes programmed to achieve a common goal by communicating through a communication medium.

Reaching consensus in a distributed system can be challenging due to communication issues and Byzantine [3] partitions. These can result in different nodes having differing views of the data, and agreeing on the exactness of the data is a challenging task. The presence of malicious nodes in the network can further complicated by attempting to disrupt the consensus process or deceive other nodes for their benefit. Designing a consensus algorithm that is both efficient and secure is a significant challenge. Despite these difficulties, DLT systems offer various potential benefits.

1.1 Advantages and Limitations of DLT Systems

The distinct characteristics of the DLT system are its redundancy and resilience. These characteristics eliminate the single-point failure with the participation of validators. The validators can validate and provide the same service or both. DLT systems can also be faster than single computer systems, i.e. scaling through to improve the performance of systems by expanding or increasing hardware resources. A fundamental problem in DLT systems is to make nodes agree on a given data, known as the *consensus problem*. Traditional DLTs rely on Byzantine Fault Tolerance (BFT) protocols to reach consensus. Typically, the BFT protocol is designed to tolerate 51% of malicious nodes[1] and works under weak synchronous network assumption. However, in digital space, it fails to solve the *double-spending* problem [4, 5]. Improved versions of BFT, Practical Byzantine Fault Tolerance (PBFT) [6] and its derivatives [7, 8] focus on providing a practical Byzantine state machine by a leader, or group of entities to address the double-spending problem. While there are several implementations in existence the most prominent implementation is blockchain technology. Blockchain technology can maintain state changes in a permission-less environment securely. Several principles are combined in blockchain technology to enable value exchange in a DLT system and solve the *double spending* and *consensus* problems.

The limitations of the DLT system are expressed using *CAP theorem* [9]. The CAP theorem defines three properties (i) *Consistency*: ensures all nodes have the identical copy of data; (ii) *Availability*: ensures that the system is accessible all the time; and (iii) *Partition tolerance*: ensures the system continues to operate during the failure of some node or links.

Distributed systems are prone to network failures generally caused by a partition; therefore, the choice is between consistency and availability in the presence of a

[1] Nodes failing or propagating incorrect information to other peers.

partition. When choosing consistency over availability, the system fails if the information is not up to date. In contrast, when choosing availability over consistency, the system will always process the query and try to return the most recent version of the information, even if it cannot guarantee it is up-to-date due to network partitioning. In the absence of partitioning, availability and consistency can be guaranteed.

1.2 Blockchain-Based DLT Systems

Blockchain-based systems have distinct characteristics that set them apart from other DLT systems. Unlike traditional databases, they bundle transactions into blocks that are cryptographically linked in chronological order to form a chain of data. The incentive mechanism encourages nodes to agree on the same chain when new and valid blocks are added. A blockchain is a novel approach in DLT to improve business processes. Through decentralization among the participants, it promotes trust via secure stores of valid transaction data in a reliable manner.

The first generation blockchain, often called Blockchain 1.0, known as Bitcoin [10] is a peer-to-peer electronic cryptocurrency system. Bitcoin proved the possibility of trust-less peer-to-peer transactions on a public network without a trusted third party [11] involvement. However, it is limited to cryptocurrency-based applications. The second generation blockchain, Blockchain 2.0, was developed as a programmable transaction platform using smart contracts [12]. This capability opened up new opportunities, such as deploying contracts on the blockchain and autonomously triggering transactions when key conditions are met. The combination of security and immutability provides an integrated level of integrity to the smart contract.

2 Introduction to Blockchain Technology

The concept of blockchain technology was introduced in October 2008 as part of a proposal for *Bitcoin*, which is a decentralised cryptocurrency system. A paper titled *Bitcoin: A Peer-to-Peer Electronic Cash System*, written under the alias of Satoshi Nakamoto, was published and later used as a reference for the Bitcoin implementation [10]. This technology enables decentralisation, transparency, immutability, and traceability for user activities. These features have significant business benefits, including greater transparency, enhanced privacy, improved traceability, increased efficiency, and faster transactions with reduced costs. The core elements of blockchain technology are blocks and transactions. Block is a holder of ordered transaction records chained with the direct neighbour block using the hash. Hash is a unique identifier of the block used to connect the new block with the previous one. The transaction is the heart of the ledger and contains the history of events that occurred between the participants in the blockchain network. Bitcoin and Ethereum are the two well-known blockchain networks. Based on the Nakamoto's paper [10] Bitcoin is a virtual currency system that issues or transfers currency without a central authority. Unlike the traditional system, it operates on a decentralised network of computers, which are purely driven by software protocols. The data is cryptographically protected and replicated among this network of computers. Each new update will go through a verification process before it reaches a consensus. The digital currency *Bitcoin* is stored in wallet software and controlled by the private and public keys of the user. There are two types of participants in the Bitcoin application: users who transact (send and receive) the digital currency and miners who process transactions through mining. Mining involves verifying transactions and solving a mathematical puzzle to create a block.

The core of the Bitcoin system is a decentralized financial network of peer-to-peer nodes that validate transactions directly between users without the involvement of third parties. When a node initiates the transaction in the Bitcoin network, it is broadcast to the other nodes in the network and is verified. After the verification, it is recorded in a publicly distributed ledger.

2.1 The Architecture of a Blockchain-Based System

The architecture of blockchain is diverse and can be explained from various perspectives. The original blockchain was identified as a public, decentralised, permissionless system. The blockchain architecture is described as a distributed append-only database, which functions as a single system on a public network of linked computers. These systems can perform transactions between non-trusted parties without going through a trusted intermediary.

In a blockchain-based system, each record of transactions is maintained across several computers are connected via a peer-to-peer network. Unlike other data structures, the blockchain's data is segmented into blocks, which are cryptographically interconnected in chronological order. These blocks are append-only data structures, which means the transactions added to the block are never removed or updated. For example, if *Bob* owns 10 bitcoins that he received from a single transaction in Block # n, this is recorded on the ledger. If *Bob* transfers 5 bitcoins to *Alice*, that will be a new transaction. In the new transaction, in Block # $n + 1$, Bob needs to use the entire output (10 bitcoins) of the previous transaction as an input. As shown in Fig. 1, Bob specifies two outputs: one refers to the transfer of 5 bitcoins to *Alice*, and the other refers to 5 bitcoins back into his account. Then *Alice* needs to follow the same way if she wants to transfer her 5 bitcoins. However, she can only spend from her received inputs. In a blockchain, *accounts* are like the placeholder of stateful objects with unique addresses. The blockchain system will keep track of the *state* of these unique addresses. Public and private key pairs are the controllers of the accounts. Bitcoin's *state* is represented by the global collection of unspent transaction output (UTXO). In contrast, the balance represents the state in Ethereum.

The data recorded on the blockchain ledger is a history of transactions which starts with the genesis block. Each subsequent block contains all transactions that

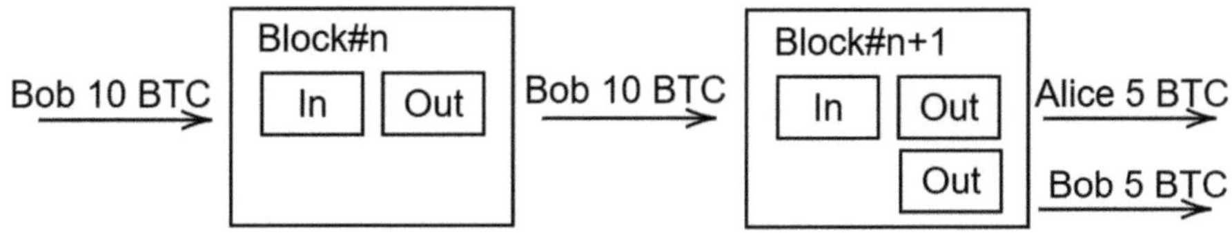

Fig. 1 An example of a Bitcoin transaction

have taken place in the network and refer to the previous block in the chain. This creates a comprehensive and continuous record of all transactions.

The transaction was initiated and broadcast to the entire node in the peer-to-peer network. The nodes then verify the validity of the transaction. Once it is validated successfully, it will include in the block. The new block then broadcasts to the entire nodes. The nodes validate the block, and once the majority of nodes agree on a new block, it becomes part of the main blockchain. In this way, blockchain ensures transparency and tamper-proof of the recorded data.

2.2 Bitcoin Value Transfer Protocol

In systems based on blockchain technology, transactions are the only atomic event allowed by the underlying protocol to update the state of them [13]. The Bitcoin application facilitates the user to update the state by extending through a new block. All successful transactions are recorded in the state of the chain. Each transaction can be reviewed anytime but cannot be updated without a collective effort. A new transaction entry must refer to its previous transaction that everybody has agreed to in which the user now transfers itself to the recipient of a previous transfer [14, 15]. Figure 2 shows the main concept of a Bitcoin transaction where each input relates to a previous output, and each output waits as an Unspent Transaction Output (UTXO) until a later transaction spends it. The genesis block and the block mining reward (Coinbase transaction) are the only exception for this process.

Bitcoin is a UTXO-based transaction system. The transaction in Bitcoin involves transferring a certain amount of Bitcoin from one participant to another. Each node in the Bitcoin network has its public and private keys. The Bitcoin address is mathematically derived from the public key. This address is used as the designation for the transaction, and the private key is for unlocking the account.

For example, when Bob wants to send bitcoins to Alice, he must refer to the block where the transaction corresponds with his public key and is signed using his private key. After that, he verifies the ownership and authorises the transfer to Alice's public address. In each transaction process, the sum of input transaction amount must be balanced with the amount of output. If the amount Bob sends to Alice is less than the input value, Bob must create a new transaction for the remaining bitcoins and

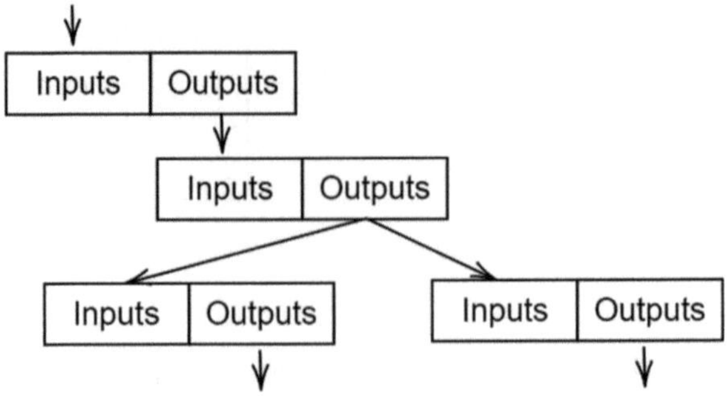

Fig. 2 Relationship between inputs and outputs in Bitcoin transactions

send them back to his account. In the blockchain, this is known as a change account; otherwise, the miner automatically accumulates the remaining bitcoins.

Blockchain operates under the rigid assumption of decentralised consensus. Mining is extending the chain by proposing new blocks through consensus. Nodes confirm transactions by mining blocks of transactions through a defined way[2] that is a function of the transactions with the history so far. More precisely, each participant maintains their local history of block records. Each block b consists of a tuple $(H(b-1), m, n)$:

- where $H(b-1)$ is a hash pointer to the previous block (b-1) in the chain,
- m is the hash of transactions contained in the block,
- n is the solution to the problem.

In Bitcoin, mining involves verifying and incorporating new blocks into the blockchain. It is performed using Proof of Work (PoW) protocol. Miners employ special hardware to solve PoW mathematical puzzles, and they receive cryptocurrency as a reward for their efforts.

The theory of the Proof of Work (PoW) protocol is described as follows in the Bitcoin white paper [10]: From the pool of transactions, validator nodes (*miner*s) try to assemble a block by combining transactions with an unknown hash puzzle. The first miner to solve the puzzle is rewarded with a certain number of cryptocurrency tokens. Solving puzzles follows a Brute-force approach, which is computationally expensive. The puzzle-solving begins when a miner receives new blocks from another miner. After that, the block's transactions verify, and the validity of the puzzle's also checked. Once the validation is successful, the chain will extend with that block and tries to solve the new puzzle with another set of transactions. The scheme elegantly justifies that transactions are validated by those willing to do so by putting some competitive effort into an incentive. This competition enables the property of security within the blockchain network. Additionally, an incentive mechanism [16] attracts miners to support the network. The miner who successfully creates a new block is

[2] A series of specialized math problems or state a token.

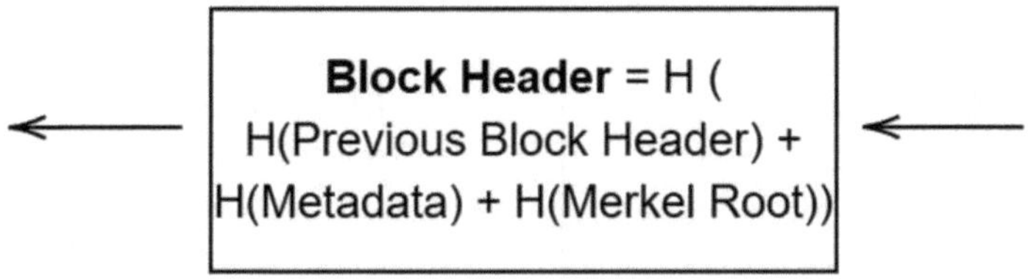

Fig. 3 Blockchain header components

granted a block reward. In general, block mining rewards and transaction fees [17] are two incentives provided by the blockchain for miners.

A block mining reward transaction is a special transaction in which a miner includes a newly mined block. This transaction credits the miner with the value of the mining reward once the network accepts the block. As per the protocol, from that point onwards, the block mining reward transaction acts as the root reference of the value created by that transaction [18]. It is created from thin air, there is no previous reference to it. Even though it does not have any previous reference, the system and its users will fully accept the value derived from it.

2.3 Blockchain System Key Components

The blockchain is made up of a chain of blocks. Each block contains a unique hash of the previous block, timestamp (referred as metadata), and transaction information (referred as Merkel Root) as major components (refer Fig. 3). These components are recorded and added to the blockchain as a block.

Block. In blockchain technology, the database is a collection of transaction data compiled and organised in a structured way called *block*. Each block is chronologically connected with other blocks using the hash. The connection of the blocks forms a chain with the collection of the transaction history. Blocks are identified by the hash of their contents available at the *block's header*. Block header is made up using the metadata,[3] a reference of the previous block, and a root hash of all the transactions available in the block (refer Fig. 4). A block header is critical information about any block, representing every bit of information about the block.

[3] Information about the block.

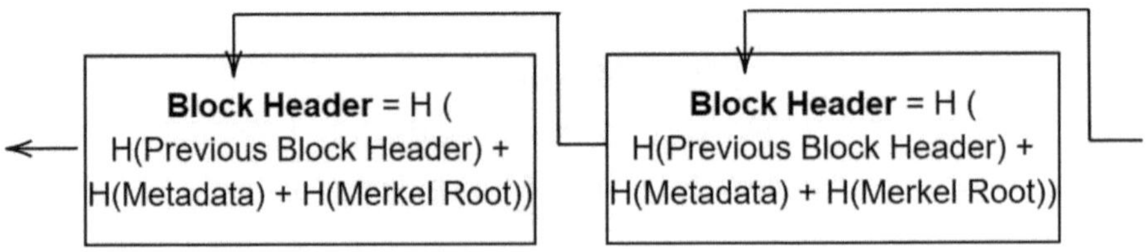

Fig. 4 Blocks forming a chain using block-hash

Hash functions. A hash function [19] is a mathematical tool that takes in numerical data input and outputs a digital fingerprint. The digital fingerprint represents the input data and is used to confirm the accuracy of the data. The ideal properties of a hash function include the following:

- Collision resistance: it is not computationally feasible to find two input messages with the same hash output where x and y are two different sets of data and $x \neq y$ and $H(x) = H(y)$.
- One-way function: for any given hash value h, it is computationally infeasible to find y such that $H(y) = h$. Therefore, it is practically infeasible to generate the input message given a hash value.
- Pre-image resistance: meaning that given a value h, it is not feasible except with negligible probability to find a value x such that $H(x) = h$.

The hash function is used in a blockchain to create a digital fingerprint of each block's data. It ensures the integrity of the contents in the block. The hash is considered a unique identifier of the block, which is available in the block header. The block header contains three components: (1) the metadata of the block, which includes the difficulty level, timestamp and nonce; (2) the hash of the previews block and (3) the transaction ID (Merkle root hash of the transactions). The genesis block has a unique structure for its header.

Each block has a reference to the previous block identity. This forms a chain of blocks and establishes the link between blocks which is cryptography secured by their hash values.

The Merkle tree structure. The Merkle tree hash technique is a way of hashing pieces of data together like a tree structure (refer Fig. 5). To build the transaction tree, two transactions are concatenated and hashed. The result is again concatenated and hashed until there is only one hash left. Each block contains a Merkle root summary of all the transactions included in that block.

The structure of the Merkle tree provides an efficient method for verifying large data sets. When a block is added to the blockchain, the hash is calculated and incorporated into the Merkle tree. The hash of the entire tree is then updated and added to the blockchain. To check the authenticity of a data block, a user only needs to obtain the block and its corresponding Merkle tree, and then compare the calculated hash of the data block with the hash stored in the Merkle tree. If two hashes match, the user can trust that the data of the block has not been altered.

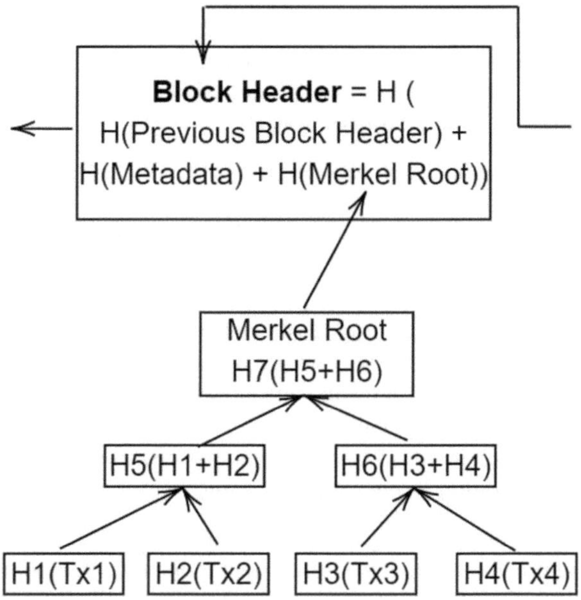

Fig. 5 A block with Merkle tree structure

The Merkle root hash for a transaction is created using a Merkle tree hash function that encompasses the hash of all transactions in the block. Any modification to a single bit in these transactions will result in a different hash value. Hence, the transaction ID serves as tamper-proof evidence of each transaction in a block, which is then included in the block header. Therefore, after a block is approved and accepted, a small change will result in a different hash value.

The network. Blockchain operates within a decentralized network, managed by a network of nodes responsible for the creation, verification, and validation of transactions. The network incentives nodes to self-govern the system by adhering to the protocol. As a technology of distributed databases, each node in the network holds a copy of the database and any updates must be shared across the network. As a result, blockchain is a decentralized data structure that maintains a continuously growing list of records, shared among all participants in the distributed network [20].

Consensus mechanism. The consensus mechanism is a critical aspect of blockchain technology that ensures the nodes in the network agree on the global order and state of transactions. This is a security measure to avoid double-spending and secure the system. Unlike traditional databases, blockchain transactions are never altered. Instead, they are recorded and kept as a permanent record, with new transactions added to the next block. Transactions are validated, grouped into blocks, and added to the chain in the order they are received. Each block is timestamped and has a unique hash linked to the previous block, forming a chain of blocks. This makes it practically impossible to modify the data once it is recorded on the blockchain.

Different consensus protocols exist, with Bitcoin using a public, permissionless PoW protocol. Private blockchains, on the other hand, operate on a permisssioned network where the identity of nodes is known and trusted, so they do not have to perform computational work to validate transactions.

Mining. Mining is a competition to solve a puzzle, where the winner is the miner who can broadcast the new block to the whole network. The sender's node does not need to rely on the other nodes on the network as long as it uses multiple nodes to ensure the transaction is spread. Conversely, the nodes on the network do not need to trust the sender, as transactions are signed and can be verified by anyone. Mining nodes perform the process according to the consensus protocol, with Bitcoin using the PoW consensus mechanism. This involves finding a number that, when hashed with the block's data, produces a result starting with a certain number of zeros. This requires investing in computing power to solve a difficult mathematical problem. The level of difficulty, which increases with the number of zeros required, is determined by the system configuration as part of the consensus mechanism.

Mining difficulty. Making mining computationally expensive, each node participating in the mining process has to invest in computation power to find the right random number (nonce) as part of the process to form the next block. This process involves finding the block header, which will be a root hash value of all the components in that block with a random number. The network requires the block header to start with the number of leading zeroes that the network defined. This process is called Proof of Work (PoW). The only way to find this random number is to perform Brute force to try the node that solves this problem to get to all the blocks of the existing chain. It is computationally hard to perform this operation but easy to verify and any other node can easily verify the result.

There are several types of consensus mechanisms, such as Proof of Work (PoW), Proof of Stake (PoS), Delegated Proof of Stake (DPoS), Proof of Activity (PoA), Proof of Authority (PoA), Proof of Burn (PoB), Proof of Space (PoSpace), Proof of Elapsed Time (PoET), Proof of History (PoH), Proof of Importance (PoI), that can be used to validate transactions in a blockchain network [21]. Each consensus technique has a unique set of benefits and drawbacks. While new algorithms are constantly emerging, PoW and PoS are the most commonly used algorithms.

Proof of Work (PoW). In the PoW network, to create a new block, a random miner must be chosen for the job. Network miners compete against each other using computational power to solve highly complex mathematical problems for the right to create a new block in the network; hence it is named Proof of Work (PoW) [22]. Additionally, the victorious miner receives a *block reward*, a fixed sum of digital money more commonly known as cryptocurrencies. Although it has been the first ever designed consensus method, it is still in use by cryptocurrencies such as Bitcoin, Dogecoin, Monero, and Litecoin. Nevertheless, this method has several scaling difficulties with high running expenses since producing new blocks demands a significant amount of computer power and energy [23].

Proof of Stake (PoS). Unlike PoW, PoS is an eco-friendly alternative that utilizes staking procedures. Users must *stake* their cryptocurrencies to be eligible to be selected as a miner on the network [22]. The more coins are staked more chances the user has to be chosen for each round. Even though this consensus mechanism is a more sustainable alternative to PoW. The system based on PoS is more favourable for those who stake the highest number of tokens, hence more likely to centralized power, which violates the blockchain principles [21].

Transactions in the Bitcoin blockchain are secure because the network employs sophisticated security algorithms, and the transaction ledger is distributed across a network of unrelated computers. To compromise the security of the Bitcoin blockchain, a hacker would need more computing power than half the nodes in the Bitcoin blockchain. Due to the size of the network participants in the Bitcoin blockchain, this is very much impossible for attacks to take over the network.

3 Technical Limitation of Blockchain Technology

The *scalability* and *interoperability* are the two major technical limitations of blockchain technology. When the number of transactions increases in the blockchain, the network can become congested, resulting in slow transaction times and higher fees. Due to this, the amount of transactions in the blocks is limited to a specific value. This brings dissatisfaction among the blockchain-based applications, which process large volumes of transactions. The second limitation is a lack of interoperability between different blockchains, which limits the potential for blockchain to be used as a universal payment system.

3.1 Scalability

In the context of blockchain, scalability is considered as a major challenge for the blockchain system. It is defined as the ability of a blockchain system to sustain performance while growing and expanding [24]. In the research work [25], scalability is identified as a major technical limitation for the wide adaptation of this technology. However, in practice, different blockchains have limitations on performance and scalability [26]. In public blockchains, scalability is considered a most important limitation, and many efforts have been undertaken to address it. Moreover, scalability is not determined by a single property in a blockchain system.

The design of permissionless blockchain systems can only process a fraction of transactions in comparison to centralised services like VISA and PayPal [27]. Due to the limitation in the number of transactions per block, a mining node can only include a relative number of transactions. Therefore, with the current architecture (until 2022) based on *Bitcoin* or *Ethereum*, a blockchain system with a given protocol can only perform a certain number of transactions.

The design of the number of transaction per second is an important factor in the blockchain system [16]. In general, the parameters *block size* and the *latency* determine the number of transactions per second. The *block size* defines the number of transactions that can be accumulated in a block, and the *latency*, sets the interval for the block creation time. The creation time is a part of the security mechanism in blockchain design.

Block size. Increasing the block size is one approach to addressing the scalability challenge in blockchain technology. The main benefit of increasing the block size limit is that it allows the Bitcoin network to process more transactions per second. This is important to bring useful Bitcoins as a part of the payment network. However, there are some trade-offs to increasing the limit of block size. For example, larger blocks mean that each individual node in the network needs to process more data. This could lead to fewer nodes being able to participate in the network, which could centralize power and make the network less secure.

As of 2022, the protocol that runs by Bitcoin has a fixed block size limit, this will limit the number of transactions that can be included in that block. The block size limit is set in regard to network propagation time. Whereas in Ethereum, the block size limit is based on the block gas limit, each transaction costs a certain amount of *gas*, and each operation is assigned a fixed amount of gas. In effect, the gas limit of a block is determined by the number of transactions that fit in a block.

Latency. Blockchain protocol has a latency or network propagation time parameter added to its protocol. Latency, the set interval for block creation time, is the time for the network to agree on standardizing the block, which belongs to the security mechanism of the blockchain design. The Bitcoin blockchain is designed to take 10 min to confirm a transaction. Each new block announcement is broadcast to the entire network [27]. This is implemented as a solution for the double-spending problem, where one can spend the money twice before the first request reaches all the nodes. So, imposing a delay will ensure that all nodes have been updated on the status before moving on to the next transaction.

In a blockchain system, the nodes are spread across the network worldwide. For effective performance, all nodes should maintain the information up to date. However, there may be network delay and latency in the network itself. As a result, there is a probability of two nodes forming a block around the same time. Both are technically valid blocks and may accept as the next block by the network. However, eventually, the network that comes up with the next block will be accepted as the longest chain by the network. This will result in rejecting the other block created even though it contains valid transactions. Discarding a block will not only be a waste of effort but will also create confusion and conflicts among the networks resulting in forking, at least for a couple of blocks. This can be avoided by imposing a block propagation time [20].

Definition 2 (*Latency*) Network latency is defined as the time required for network communication. Bitcoin protocol imposes a delay so that the network takes an average time of 10 min for a block to confirm.

To understand the creation and imposition of delay, it is important to know about two key factors. One factor is the network propagation time, and the other is the core of the mining process. The mining process involves miners solving a computational problem to form a valid block, which also acts as a security system to prove the authenticity of miners by considering their computational power. However, this is only necessary for a permissionless network. In contrast, in a permissioned network, the nodes' identities are known and trusted. Despite this, the network propagation time delay is still necessary to ensure that the information spreads across the network and validating by miners. To maintain its core properties, a blockchain system should limit its scalability. Currently, blockchain has limited scalability. In a distributed system, there is no ideal consensus protocol and a trade-off must be made between consistency, availability, and partition fault tolerance (CAP) [9]. This is similar to the CAP theorem, which states that a distributed computer system can only provide two out of the three guarantees of consistency, availability, and partition tolerance at once.

CAP theorem. In a distributed system, network failures can often result in partitions, leading to a trade-off between consistency and availability. If consistency is prioritized, the system will return an error if the information is not up to date. However, if availability is prioritized, the system will still process the request and attempt to provide the most recent version of the information, even if it may not be completely up-to-date due to network partitioning. If there is no partitioning, both consistency and availability can be ensured.

3.2 Scalability Trilemma

The trilemma in the context of blockchain states that the system can only balance the properties of decentralization, security and scalability. This is a design constraint that blockchains may not be able to solve without a trade-off.

- Scalability refers to the number of transaction processing capabilities.
- Decentralization means that every node on the network processes every transaction and maintains a copy of the entire state.
- The decentralization consensus mechanism achieves security.

In the case of blockchain, this trilemma in Fig. 6 states that decentralised blockchain networks can only provide two of three benefits at any given time concerning decentralization, security, and scalability. Consistency: every node holds a current valid set of blocks. Availability: with the distributed nature, every node has a copy of data. Partition tolerance: when a network partition occurs, the system continues to operate until the partition is resolved instead of becoming unavailable or failing.

Technically, a distributed system consists of a network of independent nodes that are connected together [2]. These nodes can only communicate through messages, i.e. there is no physical or shared memory between the nodes. As illustrated in Fig. 7, a valuable property need to maintain the balance between the *event computation time* and the time of *message transmission*. The *event computation time* $t(n1)$ and $t(n2)$ is the time for a node to perform meaningful processes, and the *message transmission time* $t(b-c)$ is the time takes to communicate messages between the nodes on the network. Leslie Lamport[4] defines a distributed system as distributed if the message transmission time is not negligible to the time between events in a single process. CAP theorem has proved that a distributed system cannot achieve consistency, availability or partition tolerance simultaneously [9]. Generally distributed systems present the problem of fault tolerance, which leads to choosing the balance between consistency over availability.

The design of a system depends on certain parameters, which determine its performance based on the outputs. In a decentralized blockchain system, every node is responsible for verifying and processing every transaction, as well as keeping a

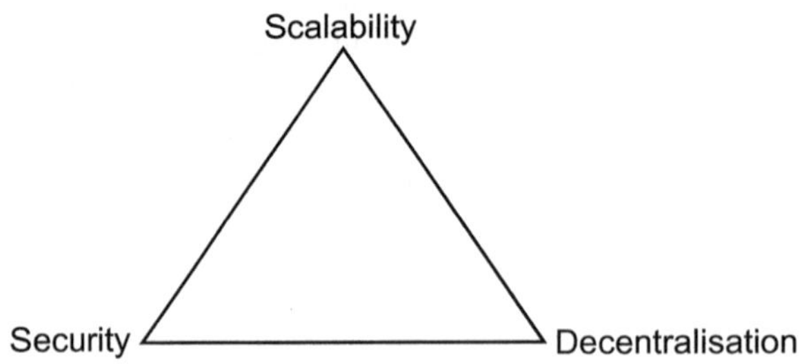

Fig. 6 Blockchain scalability trilemma

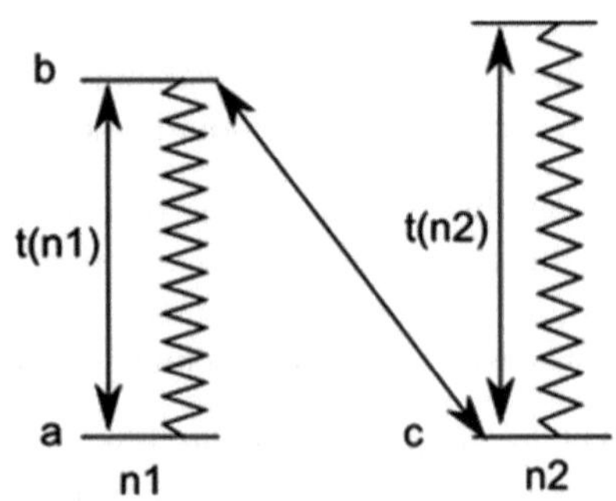

Fig. 7 Message passing between in a distributed network

[4] http://www.lamport.org.

copy of the entire system state. This design enhances the system's fault tolerance, but it also limits its scalability. For this reason, the scalability of blockchain systems must be balanced with the need to maintain their core security features. To gain a deeper understanding of the performance and limitations of blockchain technology, it is important to examine factors such as throughput and latency in relation to scalability.

In the case of blockchain, the trilemma in Fig. 6 states that a decentralised blockchain network can only provide two of the three benefits at any given time concerning decentralization, security, and scalability. An improvement in scalability to increase performance will impact either decentralisation or security.

From another angle, to address scalability and improve efficiency, a decentralised network can adopt parallel computing. In a typical blockchain system, every node has to store an up-to-date state of every transaction and perform the same operation to participate in the network. More nodes verifying the transactions will be good for the system's security but limit the scalability to a point where a blockchain system cannot process more transactions than a single node. Logically a network with thousands of nodes will have higher throughput than a single node, but a blockchain network is incapable of it. In permissioned blockchain, the design permits configuring a network of nodes to scale independently. The nodes need to have permission to operate. Thus, it removes the computation task, or even parallel computing can be possible. In contrast, in a public blockchain with Proof of Work (PoW) like a consensus, a scalability limit is built into it. A protocol with a shorter block time latency needs more confirmation for the same level of security as a protocol with a longer block time [28].

3.3 *Classification of Scalability Approaches*

Blockchain scalability approaches can be classified into three main categories:

- On-chain solutions: This refers to increasing the capacity of a blockchain network by making changes to its underlying protocol.
- Off-chain solutions: This refers to moving transactions and computations of the blockchain and into secondary networks or layer two solutions while maintaining the main blockchain security and trust.
- Interoperability blockchain networks: A solution to improve the scalability of blockchain networks by making them interoperable, allowing transactions to occur across multiple networks and reducing the strain on any network.

Each approach to scalability has its trade-offs in terms of security, decentralisation, and speed, so it is important to carefully evaluate the specific needs and requirements of a given use case before choosing the most appropriate scalability solution. Researchers are still exploring the technology and identifying the benefits of this approach.

On-chain. On-chain approaches aim to enhance blockchain scalability by adjusting the internal parameters of a blockchain. This can be achieved by improving the network latency, specifically by optimizing the transaction or block size. Here are some examples of on-chain scalability solutions:

- Sharding: is a technique that involves dividing a blockchain network into smaller sub-networks called shards. Each shard can process transactions independently, which can increase the transaction throughput of the network.
- Sidechains: separate blockchain networks connected to the main blockchain network. They allow for the creation of custom blockchain applications that can interact with the main blockchain network.

Off-chain. A recent development with blockchain scalability is the *Rollup* solutions. Rollup solutions refer to the execution of transactions outside the main network, so-called layer two networks, while anchoring the transaction data proof on the main chain, referred to as layer one. This will allow the transaction to have the security features of layer one even though the executing operations are outside of the network on layer two. There are several rollup solutions proposed to help scale a blockchain application by handling transactions off layer one while taking advantage of the robust decentralised security model of layer one. Rollup are deployed using a set of smart contracts on layer one, which is responsible for processing deposits, withdrawals, and verifying proofs. When verifying proofs, this brings the main distinction between different types of rollup, such as *Optimistic rollup* use fraud proofs and *ZK rollup* use validity proofs.

3.4 Interoperability

In information systems, interoperability is generally understood as the ability of two or more systems to communicate and exchange information [29]. The process related to exchange is the use of the exchanged data to carry out an operation, referred to as interoperation [30]. Interoperability can also be characterised as a relationship between the exchange and cooperative use of data. Interoperability occurs when two systems successfully use the exchanged data despite differences in language, interface, and execution platform [31]. Recently, interoperability has gained different definitions within the context of blockchain. Increasingly common usage refers to the *interaction* and *exchange* of data between networks of blockchains. This opens up various possibilities for cross-blockchain transactions, for example, value transfer in the form of asset or payment versus payment and payment versus delivery schemes or information exchange [32].

Generally, integration involves middleware mechanisms to transport the data and make it accessible to the other network. In the case of blockchain-based systems, the most significant obstacles to interoperability are the consensus of each chain and how data moves from one chain to another. Cross-blockchain technology has become a topic of discussion to address the integration issues of blockchain networks [33].

Blockchain interoperability solutions are designed to enable different blockchain networks to communicate and exchange information. Interoperability is becoming increasingly important as more blockchain networks emerge, each with its own features and capabilities. Here are some examples of blockchain interoperability solutions:

Third party. A third-party mode aims to provide interoperability through a trusted entity [34]. This is considered the most comprehensive method of connecting two blockchains. The scheme depends on a third-party federation of validators or notaries to attest events on another network [32]. The federation helps the destination network to verify the event that took place on the source network. In other words, the notary scheme enables the mapping of data that exist in two different networks, which helps the users to process the trade of assets.

Bridge. Bridge mode can be viewed as built-in integrated links for blockchain interoperability [34]. The link will act as a connector for blockchain network participants to access external data. A bridge system uses gateway nodes to facilitate cross-blockchain communication.

Gateway nodes interact with connected networks and perform computation based on such interactions. Depending on the distributed ledger, bridges may be full nodes resilient to crashes [35] and participate in the consensus process of the connected networks. Gateways can be embedded in the user's wallet or in a third-party server that connects the user's request to the corresponding blockchain network. Generally, a gateway is used to leverage information from the other system with which the user is not associated [36].

Gateways are also helpful in connecting blockchain nodes to the outside world to get data or information. For example, smart contracts deployed on a blockchain system usually depend on the data from their network to execute the transaction. However, there are use cases where a smart contract needs to access information outside its network. In a distributed environment, this will be extremely difficult, and miners will have different results. Gateways can help the nodes to access the data from an external source and thus can expand blockchain systems capabilities [37, 38].

Connector. The architecture of a connector mode can be seen as different blockchain networks connected through an integration hub that creates a network-of-networks [34]. The integration hub consists of a network that helps to connect the networks in a decentralised way. The hub will create the pathway for communication between network components and will have the rules that govern those interactions. The hub will facilitate the connection of many networks and act as a routing device for participating networks. The connected networks are either pre-configured as side chains with the hub or connected through a bridge in this integration mode.

These are just a few examples of scalability and interoperability solutions for blockchain networks. As blockchain technology continues to evolve, new solutions will likely be developed to address the challenges of scalability and interoperability.

4 Conclusion

Irrespective of differing views about cryptocurrency, business applications triggered significant interest in blockchain technology. In future, blockchain technology will transform into various business applications to provide its added benefits of security, transparency and accountability. The ability to create new business models makes it an exciting and promising technology worth exploring. However, as the business application evolves to develop new and innovative applications through the development of different variations and implementations of blockchain architecture. Modifications to the core design compromise the principles and fundamentally alter the nature of the blockchain and potentially undermine its stability and trustworthiness. Therefore, it is essential to approach any modifications cautiously and consider the potential consequences thoroughly. This chapter provides a brief overview of the fundamentals of DLT, Blockchain and the limitations in its design.

References

1. Natarajan H, Krause S, Gradstein H (2017) Distributed ledger technology and blockchain
2. Cachin C, Guerraoui R, Rodrigues L (2011) Introduction to reliable and secure distributed programming. Springer Science & Business Media
3. Shrivastava S, Sharma A (2022) Distributed ledger technology (DLT) and Byzantine fault tolerance in blockchain. In: Soft computing: theories and applications: proceedings of SoCTA 2021. Springer, pp 971–981
4. Chohan UW (2017) The double spending problem and cryptocurrencies. Available at SSRN 3090174
5. Hoepman J-H (2007) Distributed double spending prevention. In: International workshop on security protocols. Springer, pp 152–165
6. Castro M, Liskov B et al (1999) Practical Byzantine fault tolerance. In: OsDI, vol 99, pp 173–186
7. Bessani A, Sousa J, Alchieri EEP (2014) State machine replication for the masses with BFT-smart. In: 2014 44th annual IEEE/IFIP international conference on dependable systems and networks. IEEE, pp 355–362
8. Kotla R, Alvisi L, Dahlin M, Clement A, Wong E (2007) Zyzzyva: speculative Byzantine fault tolerance. In: Proceedings of twenty-first ACM SIGOPS symposium on operating systems principles, pp 45–58
9. Gilbert S, Lynch N (2002) Brewer's conjecture and the feasibility of consistent, available, partition-tolerant web services. ACM SIGACT News 33(2):51–59
10. Nakamoto S (2008) A peer-to-peer electronic cash system. Available at https://bitcoin.org/bitcoin.pdf. Accessed 28 Feb 2020
11. Werbach K (2019) Summary: blockchain, the rise of trustless trust?
12. Li X, Jiang P, Chen T, Luo X, Wen Q (2020) A survey on the security of blockchain systems. Future Gener Comput Syst 107:841–853
13. Burkhardt D, Werling M, Lasi H (2018) Distributed ledger. In: 2018 IEEE international conference on engineering, technology and innovation (ICE/ITMC). IEEE, pp 1–9
14. Garay J, Kiayias A, Leonardos N (2015) The bitcoin backbone protocol: analysis and applications. In: Annual international conference on the theory and applications of cryptographic techniques. Springer, pp 281–310

15. Narayanan A, Bonneau J, Felten E, Miller A, Goldfeder S (2016) Bitcoin and cryptocurrency technologies: a comprehensive introduction. Princeton University Press
16. Sompolinsky Y, Zohar A (2018) Bitcoin's underlying incentives. Commun ACM 61(3):46–53
17. Carlsten M, Kalodner H, Weinberg SM, Narayanan A (2016) On the instability of bitcoin without the block reward. In: Proceedings of the 2016 ACM SIGSAC conference on computer and communications security, pp 154–167
18. Houy N (2014) The bitcoin mining game. Available at SSRN 2407834
19. Dinh TTA, Wang J, Chen G, Liu R, Ooi BC, Tan K-L (2017) Blockbench: a framework for analyzing private blockchains. In: Proceedings of the 2017 ACM international conference on management of data, pp 1085–1100
20. Sompolinsky Y, Zohar A (2013) Accelerating bitcoin's transaction processing. Fast money grows on trees, not chains. Cryptology ePrint Arch
21. Wang Y, Cai S, Lin C, Chen Z, Wang T, Gao Z, Zhou C (2019) Study of blockchains's consensus mechanism based on credit. IEEE Access 7:10224–10231
22. Zheng Z, Xie S, Dai H, Chen X, Wang H (2017) An overview of blockchain technology: architecture, consensus, and future trends. In: 2017 IEEE international congress on big data (BigData congress). IEEE, pp 557–564
23. Tschorsch F, Scheuermann B (2016) Bitcoin and beyond: a technical survey on decentralized digital currencies. IEEE Commun Surv Tutor 18(3):2084–2123
24. Hileman G, Rauchs M (2017) 2017 global blockchain benchmarking study. Available at SSRN 3040224
25. Yli-Huumo J, Ko D, Choi S, Park S, Smolander K (2016) Where is current research on blockchain technology?—A systematic review. PLoS ONE 11(10):e0163477
26. Pillai B, Hóu Z, Biswas K, Bui V, Muthukkumarasamy V (2022) Blockchain interoperability: performance and security trade-offs. In: Proceedings of the twentieth ACM conference on embedded networked sensor systems, pp 1196–1201
27. Bonneau J, Miller A, Clark J, Narayanan A, Kroll JA, Felten EW (2015) SoK: research perspectives and challenges for bitcoin and cryptocurrencies. In: 2015 IEEE symposium on security and privacy. IEEE, pp 104–121
28. Decker C, Wattenhofer R (2015) A fast and scalable payment network with bitcoin duplex micropayment channels. In: Symposium on self-stabilizing systems. Springer, pp 3–18
29. Vernadat FB (2006) Interoperable enterprise systems: architectures and methods. IFAC Proc Vol 39(3):13–20
30. Whitman LE, Santanu D, Panetto H (2006) An enterprise model of interoperability. IFAC Proc Vol 39(3):609–614
31. Wegner P (1996) Interoperability. ACM Comput Surv (CSUR) 28(1):285–287
32. Buterin V (2016) Chain interoperability. R3 research paper
33. Pillai B, Biswas K, Hóu Z, Muthukkumarasamy V (2020) The burn-to-claim cross-blockchain asset transfer protocol. In: 2020 25th international conference on engineering of complex computer systems (ICECCS). IEEE, pp 119–124
34. Pillai B, Biswas K, Hóu Z, Muthukkumarasamy V (2022) Cross-blockchain technology: integration framework and security assumptions. IEEE Access 1
35. Belchior R, Vasconcelos A, Correia M, Hardjono T (2021) Hermes: fault-tolerant middleware for blockchain interoperability. techrxiv 14120291/1. arXiv: 1. https://doi.org/10.36227/techrxiv.14120291.V1. URL: articles/preprint/HERMES_Fault-Tolerant_Middleware_for_Blockchain_Interoperability/14120291/1
36. Hardjono T (2021) Blockchain gateways, bridges and delegated hash-locks. arXiv preprint arXiv:2102.03933
37. Hardjono T, Hargreaves M, Smith N (2020) An interoperability architecture for blockchain gateways
38. Hardjono T, Lipton A, Pentland A (2019) Toward an interoperability architecture for blockchain autonomous systems. IEEE Trans Eng Manag 67(4):1298–1309

Security Challenges and Wireless Technology Choices in IoT-Based Smart Grids

Luke Kane, Vicky Liu, Matthew McKague, and Geoffrey Walker

Abstract The Internet of Things (IoT) smart grid enables many benefits to both customers and energy generators, such as improved outage visibility, billing, and cost reduction. It allows more efficient energy use through improved access to real-time data that supports customers in reducing their energy usage and improving environmental outcomes. Integrating IoT-based data networks into the grid brings these benefits and many more. However, security and performance challenges are introduced. With the plethora of current and emerging technologies, suitable technologies must be used in each network segment that provide a sufficient level of network performance. With the introduction of data networks to the grid, we must also consider what additional threats are introduced regarding network security. This work provides essential background information on the residential smart grid. Security risks and attacks that threaten the IoT smart grid are then identified. A discussion on security challenges and future research directions are presented. A review and discussion of relevant modern IoT transmission technologies covering their benefits, key performance metrics, and their appropriate place within the IoT-based smart grid is then presented.

Keywords Internet of Things (IoT) · Smart grid · Network security · Network performance · Home Area Network (HAN) · Neighbourhood Area Network (NAN) · Wide Area Network (WAN)

L. Kane (✉) · V. Liu · M. McKague · G. Walker
Queensland University of Technology, Brisbane, QLD 4000, Australia
e-mail: le.kane@qut.edu.au

V. Liu
e-mail: v.liu@qut.edu.au

M. McKague
e-mail: matthew.mckague@qut.edu.au

G. Walker
e-mail: geoffrey.walker@qut.edu.au

L. Kane
Cyber Security Cooperative Research Centre, Brisbane, Australia

S. Pal et al. (eds.), *Emerging Smart Technologies for Critical Infrastructure*,
Smart Sensors, Measurement and Instrumentation 44,
https://doi.org/10.1007/978-3-031-29845-5_7

1 Introduction and Background of IoT-Based Smart Grid Architecture

The smart grid is a critical technological advancement that will continue to assist the world in generating and consuming electricity more sustainably. No longer is electricity generated and delivered to consumers in a one-way manner. With the two-way flow of energy and data, the smart grid enables improvements in many areas, such as monitoring grid stability, billing and metering, and remote control of appliances [1]. Internet of Things (IoT) based wireless transmission technologies are a vital part of the network infrastructure of the smart grid, with a diverse mix of them used [1].

Technology that is used to construct and deliver IoT-based networks should provide value for money by using affordable components and ideally be based on unlicensed band transmission technologies without network access fees [2]. Energy efficiency is also a key consideration as some devices may run on batteries and need to operate for extended periods without intervention [2]. Since the smart grid is critical infrastructure and transmits user data, we cannot and should not ignore network security. Data confidentiality, authentication, and integrity must be considered in any smart grid network [3].

Security measures, by their nature, can be computationally expensive and may negatively impact the expected performance of a system, device, or network [4]. This is especially true in IoT-based networks that often consist of resource-constrained devices. The fact that security can come at the expense of performance does not negate the need for solid and effective security measures to protect critical systems such as the smart grid.

In the remainder of Sect. 1, we present background information on the structure and architecture of the IoT-based smart grid communication networks. Topics that are covered in this section include the architecture of the smart grid and background information on the Home Area Network (HAN), the Neighbourhood Area Network (NAN) and the Wide Area Network (WAN). We will then discuss the three-layer IoT model, which will be relevant for categorising attacks later in the chapter. A discussion of relevant previous research in each of these networks is presented.

In Sect. 2, we will discuss some relevant previous security-related research in the HAN, NAN, and WAN and examine attacks and challenges that impact the smart grid and, more broadly, IoT networks in general. Future research directions in the IoT-smart grid are then identified. Section 3 presents a review of contemporary IoT transmission technologies that are relevant to the smart grid. These key technologies will be identified, and their strengths and weaknesses will be discussed in terms of performance and security. The data network that would benefit the most from each technology will be identified. This chapter will conclude in Sect. 4 with a summary of the key discussion points.

1.1 Smart Grids

In simple terms, the smart grid combines the traditional electricity grid with modern technology that allows not only the delivery of electricity but also bi-directional communication between energy producers and consumers [5, 6]. The smart grid consists of various systems and components that are interconnected, allowing more efficient control of the grid through events such as changes in demand and equipment failure [5].

The smart grid brings many broad benefits over the traditional grid. One such benefit is allowing energy providers to monitor the health of the grid in real time. This allows actions to be taken to reduce events such as blackouts and brownouts [7]. This action is enabled by the availability of real-time data allowing the supply to be balanced quickly in the event of grid instability [6]. The smart grid is a key enabler of efficient distributed energy generation [8]. This allows renewable energy sources such as wind and solar to be integrated into the electrical grid. As the sun is not always shining, and the wind is not always blowing, this must be done in a considered manner to maintain a stable grid [8].

Within the smart grid, there is also the concept of the microgrid. With the proliferation of distributed energy resources, we are seeing smaller, more localised generation and storage technologies [9]. In the event of disconnection from the wider grid, these microgrids can continue to deliver power to the affected customers by operating in island mode [9]. This is a further benefit of having a distributed smart grid architecture.

While the benefits and the inevitability of the smart grid are clear, the expansion and integration of communication technologies into the grid is not without its challenges. The cyber security risks cannot be understated. The smart grid is an enticing and attractive target for attacks by state-based actors and other cyber criminals [10, 11]. While the smart grid can be the primary target for an attacker, it can often be a means to disable another target or to achieve other goals by interrupting the electricity supply [11]. Security risks and attacks that can impact the IoT-based smart grid, along with mitigation strategies and future research directions, will be discussed in Sect. 2.

The communication networks of the smart grid can be classified into three broad categories. These can be defined as [12]:

1. Home Area Network (HAN)—Located inside the customer's home for appliance control and automation. The HAN will be discussed in Sect. 1.2.
2. Neighbourhood Area Network (NAN)—Primarily, the network of smart meters of different premises in a geographical neighbourhood. The NAN is responsible for data aggregation and is the link between the customer network and the energy provider. The NAN will be discussed in Sect. 1.3.
3. Wide Area Network (WAN)—This network acts as the backbone and connects the various distributed neighbourhoods to the main network and the control centre. The WAN will be discussed in Sect. 1.4.

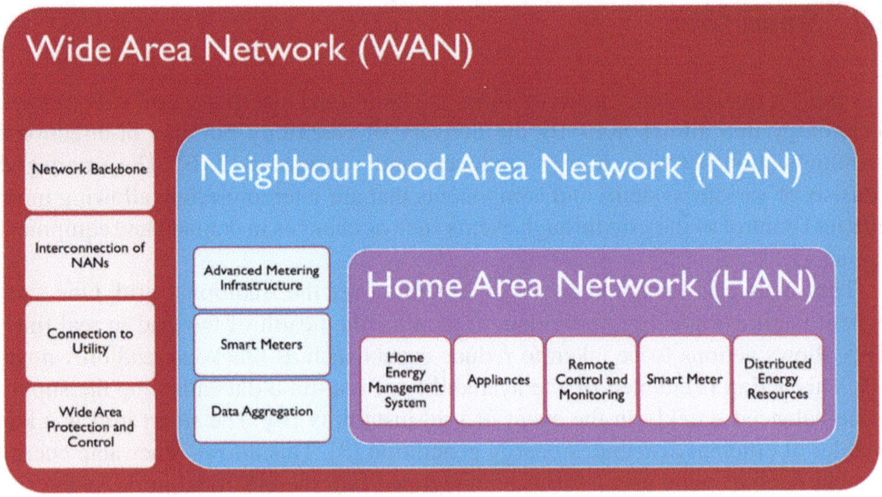

Fig. 1 The three communication networks of the IoT smart grid, including each network's key participants and functions

A visual representation of these networks and each network's key functions and participants can be seen in Fig. 1. The figure is intended to convey a hierarchical structure with the HAN communicating via the NAN to the WAN and vice versa.

Wireless communication technologies are the basis for IoT networks, including the smart grid. Relevant wireless communication technologies will be discussed in Sect. 3.

1.2 Home Area Network (HAN) and Home Energy Management Systems (HEMS)

This section will discuss the HAN and Home Energy Management System (HEMS). It will define the HAN/HEMS and discuss the benefits, structure, security concerns, and the previous work relating to HAN/HEMS secure architectures. With the declining state of the climate, it has never been more critical to reduce our reliance on fossil fuels and our energy consumption. Worldwide, residential households account for around 17% of carbon emissions [13]. This figure can be reduced with the right tools and data available for households. Users monitoring their energy consumption can promote behavioural changes by making them accountable [14].

The HAN enables systems that are used in the smart home to monitor energy usage and control appliances. These systems are known as Home Energy Management Systems (HEMS) [14–16]. HEMS gives users the ability to monitor and better control their power consumption. These systems can provide monitoring, control, and logging functions [17]. The HAN/HEMS can also facilitate appliance communication with a smart meter to make decisions on when the most appropriate operation time might be [18]. The components of the HAN and the HEMS can include distributed energy resources, appliances, networking technologies, and a control/management system [17].

More and more households are embracing renewable energy sources and are installing solar panels on the roof of their homes [19]. This is driven not only by a desire to reduce the impact on the environment but also by the continued rising energy costs that are being experienced globally [19]. The HEMS can also provide demand response price-based incentives to users, allowing the usage patterns of schedulable appliances to be altered [17, 20]. Many consumers would like to reduce their reliance on traditional energy producers, with many desiring complete self-sufficiency [19]. Household battery storage solutions will be critical to consumers achieving this goal [19].

1.3 Neighbourhood Area Network (NAN)

The NAN is the network that carries data from the various HANs to and through to the utility provider via the WAN. The NAN communication network complements the electricity distribution network [21]. It is essentially a network of smart meters in a neighbourhood connected to a data aggregation point [21]. The NAN is the backbone of the advanced metering infrastructure (AMI) system. It supports functions such as management of power outages, demand response functions, meter reading, distribution automation, and monitoring [22]. These primary functions of the AMI have relatively low data rate requirements with a maximum of around 100 kbps [22, 23]. A typical NAN may have a radius of between 1 and 10 km [24].

1.4 Wide Area Network (WAN)

The WAN complements the electricity generation and distribution electricity network and provides the link between the NAN and the utility provider [21]. The WAN can support wide-area protection and control. This enables automatic monitoring and protection of the grid to protect against and recover from events that may cause outages or instability [25]. The WAN needs to support the monitoring and control of not only the smart home and the smart neighbourhood but also the entire smart grid. Due to this fact, the transmission speed requirements of the WAN are typically much higher than required in the HAN and the NAN and can range up to 1 Gbps [25, 26]. The WAN must also support vast distances of up to 100 km, so a mix of high-speed wireless transmission technologies and wired fibre optics are typically used to provide the network infrastructure [25].

1.5 *The Three-Layer IoT Model*

IoT networks, like other networks, are described in the literature by several different models. One of the most commonly used models to describe IoT architecture is the three-layer model [27, 28]. These layers are:

- Perception/Physical Layer—The physical components that make up the IoT network. This can include hardware devices, as well as sensors and actuators.
- Network Layer—This layer describes the technologies and systems that allow communication and transmission to occur.
- Application Layer—This layer describes any software services that use the underlying network and physical/perception layers to supply some sort of functionality.

A graphical representation of the three-layer model can be seen in Fig. 2. The figure shows how the layers interact with the hardware at the bottom, servicing both the network and application layers. The IoT three-layer model is relevant, as, in Sect. 2.2, various attacks will be classified according to this model.

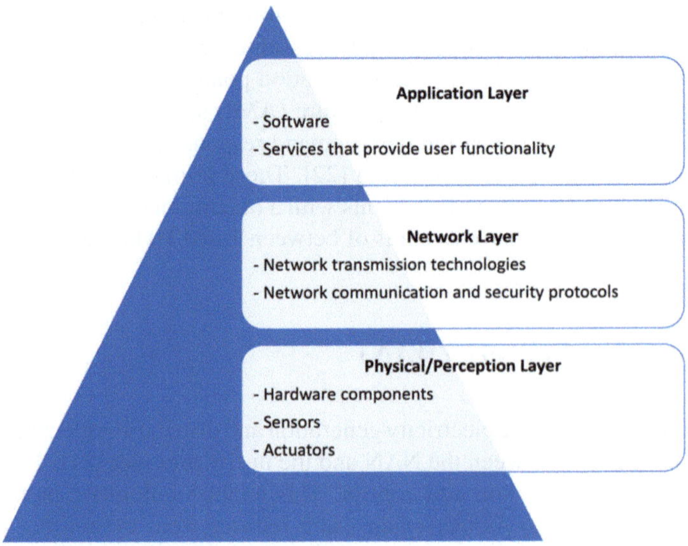

Fig. 2 A visual representation of the three-layer IoT architecture model, including the components that belong to each layer

2 Attacks and Challenges

This section will discuss the threats and challenges faced by the IoT-based smart grid. We will first discuss some previous security-related research in the context of the HAN, NAN, and WAN. Attacks relevant to the IoT-based smart grid will then be categorised based on the model, including the type of attack, a description of the attack, the security goals that are compromised, and suggested mitigation/detection strategies. We will then discuss some of the other challenges present in the IoT smart grid and outline topics that the literature has identified as key future research areas.

2.1 HAN/NAN/WAN Security-Related Research

This section will discuss some previous research relevant to the IoT-smart grid. A selection of previous research relevant to the HAN, NAN, and WAN will be presented.

The HAN and HEMS provide many benefits to all participants in the smart grid; however, by integrating new data networks into the smart grid system, the existing threats and vulnerabilities of those networks are introduced into the grid [29]. Not only does this introduce these risks into the broader electrical grid, but it also risks users' privacy through the unauthorised access and disclosure of their personal information [29]. These vulnerabilities could also threaten the stability of the power supply. The risk of a power failure can be a safety issue and a threat to life, with many people reliant on medical devices [30]. People may rely on medications for survival that are temperature sensitive, so any prolonged loss of power could impact the refrigeration of these items [30]. There is also a broader risk to personal safety and property safety. If an attacker could manipulate a home appliance into malfunction, this could cause injury or death [31]. Steps to mitigate these risks should be taken during the design phase of any HAN [29].

Menon and Radhika [32] highlighted that smart meters can communicate detailed energy usage patterns about a given consumer to the energy provider. Disclosure of this information to an unauthorised third party is a significant breach of the user's privacy. In their work, they proposed a scheme to detect anomalous network traffic that could be related to flooding attacks. Their study considered patterns that could be related to DoS attacks such as ping of death, Internet Control Message Protocol (ICMP) and User Datagram Protocol (UDP) flooding attacks. Using their algorithm, they were able to identify these anomalous traffic patterns during their evaluation.

Ali et al. [33] discussed the particular vulnerability of the HAN to attacks. Their rationale is that many consumers do not possess knowledge of cyber threats and best security practices. They determine that this makes the HAN the most vulnerable part of the smart grid. To account for the lack of user network security literacy, they state that all technologies deployed in the HAN should have built-in security measures to protect the user from attacks and privacy breaches. They propose a data aggregation scheme to forward cost information from the HAN to the utility provider. This is to mitigate against a breach of privacy occurring to users if their usage data is intercepted or obtained by an unauthorised third party.

Like the two previously discussed studies, several other recent studies also discuss anomaly and threat detection in the context of the HAN [34–36].

Like the HAN, security must also be considered in a NAN deployment. The consequences of a security breach in the NAN could be severe and include data loss, energy theft/fraud, privacy breaches, and instability of the energy infrastructure [37]. There is also the risk of man-in-the-middle (MITM) attacks, physical attacks on the smart meter devices and insider attacks on the NAN gateway devices/data aggregation points [38].

Yilmaz and Uludag [39] propose a hierarchical intrusion detection system (IDS) to mitigate the risk of a DoS or DDoS attack. Their proposal includes a general threat model and an IDS framework that operates in a distributed fashion in both the HAN and the NAN.

Kalidass et al. [40] proposed a scheme to improve privacy and data integrity within the AMI. They proposed a key agreement scheme that uses a new session key for each message sent between the smart meter and the utilities head-end system. These session keys are established from a long-term key that is generated on an authentication server. Kalidass et al. argue that due to the regular changing of the session key, confidentiality and integrity are improved in the case of a compromised session key.

Alohali et al. [41] proposed an authentication scheme to authenticate groups of smart meters with the energy provider rather than individual meter authentication. Their study is interesting in that it approaches network security in a way that promotes confidentiality and authentication with a strong focus on scalability. Their scheme works by each NAN gateway authenticating a small group of smart meters or other NAN gateways. The gateways can then authenticate with the energy provider on behalf of their group of smart meters.

Attacks that are directed toward the WAN may be targeting the power infrastructure [42]. Kharchouf et al. [43], in their work, examined the security of the WAN and, in particular, the management of wide-area measurement systems and phasor measurement units. They noted that an attack on these systems could have profound consequences and result in the utility provider making incorrect decisions about the operations of the electricity grid.

As the WAN data network supports the transmission and distribution part of the electricity network, a breach in the WAN can cause widespread problems. If an attacker gains access to the components in the WAN and can manipulate operational data such as supply and demand information, this could cause power blackouts [44].

2.2 *Attacks on IoT Networks*

The smart grid faces many cyber-security challenges. The security goals of the CIA triad (confidentiality, integrity, and availability) are just as relevant in the IoT smart grid as in any other network [10]. In addition to the CIA triad, it is vital also to consider authenticity, authorisation, and non-repudiation [45].

Many threats and attacks are common to all IoT-based networks, including the smart grid. The main difference between a generic IoT network and the smart grid is that the smart grid is critical infrastructure, so the consequences of an attack are potentially more disruptive or severe.

Liang and Kim [46] categorised attacks that can impact IoT networks using the three-layer model as a basis. A summary of some of the most relevant attacks, the affected security goals, and strategies for mitigation/detection can be seen in Table 1.

This is by no means an exhaustive list of potential attacks that occur in IoT networks. It is merely a summary of some of the more well-known attacks. These threats are just some of the challenges faced by the smart grid and other IoT-based networks. When designing or implementing a network, the designer should seek to develop strategies to mitigate these threats.

2.3 Security Challenges and Future Research Opportunities

This section will discuss some relevant security-related challenges faced in the IoT-smart grid and summarise some future research opportunities discussed in the literature.

IoT devices are by nature resource-constrained, meaning they typically have less processing ability, memory, and bandwidth than a traditional system and, depending on the device, may operate without a fixed power source (battery-powered) [51]. This introduces challenges in implementing standard security mechanisms and cryptographic algorithms [52]. Symmetric key lightweight encryption schemes can overcome some of these security challenges; however, it is essential to also use an authentication mechanism [53]. Asymmetric schemes such as RSA are generally unsuitable for IoT networks; however, ECC (Elliptic curve cryptography) based schemes may be a viable option due to their reduced key length and less resource-intensive nature [53]. It is estimated that within the next two decades, advances in quantum computing could threaten the viability of public key-based systems [54]. This fact should be considered due to the longevity of some devices in the smart grid. The current state of the research suggests symmetric key cryptography will remain a viable option; however, keys may need to increase in length [54].

On another encryption-related topic, there is also a key distribution problem. The scale of the IoT network in the smart grid is vast, with potentially many millions of devices. Efficient ways of ensuring all devices have the appropriate cryptographic keys and any other associated data is a challenge during initial network deployment and on an ongoing basis [55].

Authorisation is also of concern in the IoT-smart grid. Due to the ability for remote management and monitoring of devices, authorisation and access control frameworks are required to limit the access rights of all users [55]. Users or devices with too many unnecessary privileges increase the risk of attacks occurring either from an attacker

Table 1 Security attacks that can occur in IoT networks categorised by the three-layer IoT model [46], including which security goals have been compromised [47] and appropriate mitigation and/or detection strategies [47–50]

Layer	Attack	Attack description	Compromised security goal(s)	Mitigation/detection strategies
Physical	Reverse engineering and tampering	Figuring out how a device works to exploit its vulnerabilities or physically extracting sensitive data from a device	Confidentiality, integrity	Regular firmware integrity checking of devices
	Jamming	Creating RF interference to disrupt the connection of other devices to the network. This is a form of Denial of Service (DoS) attack	Availability	Frequency hopping
	Social engineering	Using deception to obtain sensitive information or gain access to a system	Confidentiality	Staff training and cyber security education
Network	Man-in-the-middle/eavesdropping	The capture and potential modification of data in transit. This can also involve the retransmission of a previously captured packet to manipulate the target to complete an action (replay attack)	Confidentiality, integrity	Strong authentication mechanisms, sequence numbering of packets
	Spoofing	Changing addressing information in a data packet to appear to be from a legitimate source	Integrity, availability	Strong authentication mechanisms, detection of excessive connections

(continued)

Table 1 (continued)

Layer	Attack	Attack description	Compromised security goal(s)	Mitigation/detection strategies
	Flooding	A form of Denial of Service (DoS) or Distributed Denial of Service (DDoS) attack involves flooding a target system/device with a large amount of traffic so that it becomes overwhelmed and unable to respond to legitimate requests	Availability	Network traffic filtering, limiting data rates at the router, isolation of components under attack
Application	Code injection	The insertion of additional code to trigger an unauthorised action, obtain information, or harm/damage/alter a database/data or system	Integrity	Validation of data and discarding of illogical or impossible values
	Phishing	Using deceptive tactics to trick an unsuspecting user into accessing a malicious site/system to either install malicious software or obtain sensitive data like usernames and passwords	Confidentiality	Staff training and cyber security education

or, potentially, a disgruntled employee [55]. This could lead to both user privacy issues and power stability issues, depending on the system that is compromised [55].

It is essential that the IoT devices used within the smart grid networks can accept over-the-air updates. New vulnerabilities are discovered in systems over time. Having the ability to remotely update and patch the IoT devices will help to prevent them from being compromised by attackers or used for nefarious purposes [56]. Having an over-the-air update system itself also introduces its own challenges. If the system's security features are not designed properly, it can lead to compromised IoT devices [56]. An insecure boot process can exacerbate this problem, as if an attacker does manage to compromise the device, on reboot, it could launch malicious code [57].

Another problem is the lack of standardisation across the IoT space [58]. Due to the extreme diversity in types of devices, and manufacturers, there is a somewhat fragmented landscape of devices. This also leads to a lack of interoperability between devices and concerns about maintaining secure communications [58]. There have been attempts to ensure interoperability in the smart grid, such as the NIST Framework and Roadmap for Smart Grid Interoperability Standards [59]; however, interoperability remains an ongoing challenge.

It is also common that many IoT devices have little or no auditing mechanisms, such as event-logging capabilities [57]. This is a particular problem within smart devices aimed at the HAN. It is not uncommon for cheap, inexpensive devices to be particularly vulnerable [57]. With no auditing controls, it is hard to identify and prevent any unwanted activity. Auditing is also crucial from a billing perspective. If there is a doubt in billing accuracy, there must be a means to investigate and determine if there is a problem and who or what the cause is [10]. It is also essential that any auditing scheme is paired with an effective authentication scheme [10].

The smart meters in each home are a prime target for attackers, affecting users' physical security and privacy. If an attacker gains access to the data of when energy is used or when certain appliances are used, they could use this information to plan home invasions or other thefts [60].

There are also broader privacy challenges in IoT networks. As IoT becomes more pervasive, there is an increased concern around user privacy [61]. With devices constantly listening to and watching us, there is a concern about the privacy of our data and the right to individual privacy [61]. This raises questions about the need for more robust and appropriate legislative frameworks that protect user privacy and enforce consent requirements [61].

So far in this section, we have discussed security and privacy-related challenges; however, safety-related challenges are just as significant. Compromising the IoT smart grid can also lead to genuine safety concerns. Suppose an attacker can compromise devices within the smart home. In that case, they could manipulate these devices into a dangerous situation and cause injury or death. An example would be the opening of a gas valve or manipulating an appliance into a dangerous action [31]. Other harmful actions could be taken, such as blocking signals from sensors and cameras that could otherwise identify and alert a home user of a hazard. Actions such as disrupting smoke sensors in the event of a fire could prove fatal [31].

A constant and stable grid is essential to the functioning of modern society. If grid stability and power supply are impacted, the consequences are potentially severe and widespread. Power outages can affect medical systems, causing health impacts and a potential loss of life [30]. Depending on the duration and frequency of blackouts, it can also impact the economy, and agriculture production, potentially impacting food security [30].

Meaningful research must continue in IoT smart grid security. Some critical areas for future research that have been identified in the literature are:

1. The creation of new protocols and frameworks that can support the smart grid's current and future security requirements [10].
2. Implementing effective auditing controls [10].
3. Examine the potential security issues and implications of the increasing integration of distributed energy resources [10].
4. Self-healing IoT-based solutions for networks in case of faulty or compromised devices [10, 62].
5. Unification and interoperability of the various relevant smart grid and security standards that are currently in existence [10, 62].
6. Novel ways to protect the privacy of users' data. The information obtained from smart meters and other smart home devices could reveal information about a household and their habits that could be compromised by criminals or used for unsolicited marketing [62].
7. Authentication and authorisation mechanisms that can identify legitimate network participants and restrict their available actions [62].

3 IoT Transmission Technologies for the Smart Grid

This section will discuss a selection of IoT transmission technologies that can be deployed in the smart grid. A discussion of the features and benefits of each of the following IoT transmission technologies is presented. The limitations and the appropriate area of the smart grid for which the technology is suitable are also discussed. Relevant security mechanisms are also identified and discussed. The technologies that will be addressed include LoRa/LoRaWAN, Bluetooth, Zigbee, Thread, Wi-Fi (including traditional Wi-Fi and Long-Range Wi-Fi 802.11ah), SigFox, NB-IoT, LTE-M, and 5G.

Table 2 summarises some of the key performance areas of each of the reviewed transmission technologies. These include:

- The maximum coverage range that the technology can achieve.
- That maximum data rate (throughput).
- If the technology is free to access (Unlicensed band) or has an associated cost (licensed band).
- If there are any restrictions on the transmission frequency (duty-cycle bound).

Table 2 A comparison of the key performance areas of the discussed IoT transmission technologies

Technology	Spectrum licensing	Duty cycle	Max coverage	Max throughput	Max payload	Security measures	Suitable for
LoRa sub-GHz	Unlicensed	Yes	> 10 km [63–65]	27 kbps [66]	255 bytes [67]	No built-in security	NAN, WAN
LoRaWAN	Unlicensed	Yes	> 10 km [63–65]	27 kbps [66]	243 bytes [68]	Symmetric multikey encryption system [69]	NAN, WAN
LoRa 2.4 GHz	Unlicensed	No	443 m [70]	250 kbps [71]	255 bytes [72]	No built-in security	HAN
Bluetooth	Unlicensed	No	100 m [73]	3 Mbps [74]	247 bytes [75]	AES-128 encryption [76, 77]	HAN
Zigbee 802.15.4	Unlicensed	No	100 m [78, 79]	250 kbps [80]	128 bytes [81]	AES-128 encryption [82]	HAN
Thread	Unlicensed	No	100 m [79, 83]	250 kbps [84]	128 bytes [81]	AES-128/ECC encryption [85]	HAN
Wi-Fi 802.11ax	Unlicensed	No	~ 100 m [86]	> 1 Gbps [86]	A-MPDU 4,194,304 bytes [87]	WPA2/WPA3 [88]	HAN
Wi-Fi 802.11ah	Unlicensed	Yes	~ 1 km [79, 89]	86.7 Mbps [90]	A-MPDU 7991 bytes [91]	WPA3 [90]	NAN
SigFox	Unlicensed[a]	Yes	40 km [92]	100/600 bps region dependent [2, 93]	12 bytes [93]	AES-128 based authentication, optional encryption [94]	NAN, WAN

(continued)

Table 2 (continued)

Technology	Spectrum licensing	Duty cycle	Max coverage	Max throughput	Max payload	Security measures	Suitable for
NB-IoT	Licensed band	No	10 km [95]	200 kbps [95]	1600 bytes [92]	Security architecture inherited from LTE [96]	NAN, WAN
LTE-M	Licensed band	No	5 km [97]	1 Mbps [97]	1280 bytes [98]	Security architecture inherited from LTE [96]	NAN, WAN

[a] Access fees apply to this technology

- If there are any built-in security measures.
- The appropriate smart grid network (HAN/NAN or WAN) for which the technology would be most suitable.

Further information about each of the surveyed technologies can be found throughout the remainder of Sect. 3.

3.1 LoRa

LoRa is an unlicensed band, sub-GHz proprietary Low Power Wide Area Network (LPWAN) technology developed by Semtech [99]. LoRa communicates using a Semtech proprietary physical layer implementation based on a modulation technique known as Chirp Spread Spectrum (CSS) [100]. Others are free to implement their own media access control mechanisms based on the physical layer. The LoRaWAN protocol is commonly used (but not always) with LoRa networks [99] and will be discussed in Sect. 3.2.

LoRa's maximum communication range is greater than 10 km [63–65], and the throughput can reach a maximum rate of 27 kbps [66], with a maximum payload size of 255 bytes [67]. LoRa has duty cycle restrictions in many parts of the world that limit the frequency of transmissions that can be sent in any given hour [66]. LoRa has both a significant range and robustness to noise [101].

LoRa has parameters that can be tuned to influence network performance in terms of coverage, throughput, and redundancy [102]. Increasing the bandwidth will increase the transmission rate while decreasing the communication range [103]. As LoRa is based on CSS, the spreading factor defines the chirp rate [104]. A lower spreading factor increases the chirp rate; this causes a faster transmission rate with a lower communication range [103, 104]. The chirp rate is halved with each increase in the spreading factor [104]. Increasing the code rate will introduce redundancy into the transmission, improving the resilience while increasing the transmission time [103]. The transmission power can also be tuned.

There is no security implementation built into LoRa, so additional security arrangements should be considered before implementation.

Due to its long range, it has an application, particularly in the NAN and the WAN. However, its usefulness will depend on the application requirements. Due to the duty cycle limitations, it may not be the most suitable choice for real-time transmission requirements.

3.2 LoRaWAN

LoRaWAN is a protocol maintained by the LoRa Alliance. It is a media access control protocol and network architecture that can be used in conjunction with sub-GHz LoRa. LoRaWAN has a star-of-stars topology that can use multiple gateways that act as forwarding nodes to send traffic from the LoRa network to a central network server, which converts the traffic into IP-based traffic [105]. This traffic is then processed by multiple servers such as network servers, join servers, and application servers [105]. LoRaWAN has classified devices into three distinct categories:

- Class A. This mode can save the most energy. These devices will only wake to send an uplink transmission. They can only receive a downlink message from the gateway directly after an uplink transmission has occurred [105].
- Class B. These devices use a schedule for when to wake for an incoming downlink transmission [105].
- Class C. These devices consume the most energy. They are always ready for an incoming transmission [105].

Security in LoRaWAN is provided by a symmetric multi-key design that provides confidentiality and integrity, with keys protecting network communication and separate keys protecting application-specific data [69].

LoRaWAN paired with LoRa sub-GHz has a clear application in the NAN due to its noise resilience and long-range coverage. As mentioned above, LoRa sub-GHz has a maximum range greater than 10 km when using a high spreading factor and low bandwidth configuration [63, 65, 106]. A LoRaWAN network has a maximum payload length of 243 bytes [68].

Given the architecture design of multiple gateways and backend servers, implementation into the HAN may not be practical.

3.3 LoRa 2.4 GHz

Semtech recently released the SX1280 LoRa chipsets. These chips can operate on the 2.4 GHz band [70, 107]. Like its sub GHz predecessor, it uses a CSS-based modulation and forward error correction to protect from noise and interference and to generally improve its resilience and robustness [108].

Unlike its sub GHz predecessor, it is not subject to duty cycle limitations and can provide faster data throughput, up to 250 kbps [71]. Also similar to sub-GHz LoRa, It provides the same customisable parameters spreading factor, bandwidth, code rate, and transmission power [108]. LoRa 2.4 GHz can provide coverage up to 74 m indoors and up to 443 m outdoors [70]. The maximum throughput of LoRa 2.4 GHz is 253.91 kbps [70]. Like its sub-GHz variant, LoRa 2.4 GHz supports a maximum payload size of 255 bytes [72].

LoRa 2.4 GHz with a custom MAC layer may be an appropriate choice for the HAN. The 2.4 GHz frequency is not subject to a duty cycle, so it is more suitable for real-time operations such as monitoring and controlling smart home appliances. Precautions should be taken to ensure the network is secure and has measures to protect confidentiality and integrity, as there are no built-in security features.

3.4 Bluetooth

Bluetooth is a mature short-range transmission technology that was first created in 1994 [109]. It operates in the unlicensed band at 2.4 GHz [77, 109]. The standard is maintained by the Bluetooth Special Interest Group, currently at version 5.3 [110]. Bluetooth has multiple topology options, such as point-to-point, broadcast, and mesh [76, 77]. The mesh topology is promising as this would help to extend the communications range.

Bluetooth mesh currently uses the Advanced Encryption Standard with 128-bit keys (AES-128) for encryption in the physical, network and application layers [76, 77]. Traditionally, Bluetooth may not have been considered a viable technology to underpin IoT networks. With fifth-generation Bluetooth, implementations now have increased throughput and coverage, reductions in power consumption, and the addition of meshed capabilities [77]. These improvements now make Bluetooth a viable option that warrants sincere consideration for IoT network deployments.

When looking at the usefulness of Bluetooth in smart grid deployments, the limited range and amount of supported network nodes (up to 8 devices) is a concern [73, 111]. The range is highly variable and dependent on numerous factors and can be anywhere in the range of 10–100 m [73] with a maximum throughput of 3 Mbps [74]. It supports a maximum packet length of 247 bytes; however, in practice, this is limited by the MTU of the given implementation [75]. Due to its limited range, Bluetooth is most suitable for replacing cables between traditionally wired devices such as in a personal area network configuration [73] and, as such, is most suitable for the HAN.

3.5 Zigbee

Zigbee is a short-range transmission technology that operates in unlicensed bands [112]. The standard is managed and updated by the Connectivity Standards Alliance (formerly the Zigbee Alliance). The physical layer and media access control (MAC) layers that Zigbee uses have been pre-defined by the IEEE 802.15.4 standard [113]. The Zigbee protocol is implemented in the application and network layers [82]. Zigbee supports multiple network topologies [80]. The coverage provided by Zigbee varies. At its maximum, it can provide up to 100 m of coverage [78]. This quoted range is the maximum provided by the underlying IEEE 802.15.4 transmission technology

[79]. The maximum throughput is around 250 kbps [80]. The maximum frame length of the underlying IEEE 802.15.4 technology is 128 bytes [81].

Zigbee uses AES-128 for encryption. Three factors are used to ensure network security: the network key, link key and master key [82]. Zigbee is a viable option for IoT networks; however, the coverage range is not ideal. Zigbee could be combined with long-range technology to help address this shortcoming, as discussed in [78, 114]. Due to the short range, the HAN would be the most appropriate place for Zigbee.

3.6 Thread

Thread, like Zigbee, is built on top of the IEEE 802.15.4 standard. As it relies on the physical and MAC layers from IEEE 802.15.4, this puts Thread into the category of short-range transmission technology operating in the unlicensed band. The standard is maintained by the Thread Group, with the latest version being 1.3, last updated in 2022 [115]. An open-source implementation has been released by Google [116]. An advantage of thread is that it is based on IP, so it can communicate on the Internet natively [117]. Thread operates using UDP and can optionally implement TCP if required [117, 118]. Thread networks are designed to be 'self-healing', meaning if devices are added or removed, the network can automatically adapt to these topology changes without intervention [117]. As Thread is a MAC layer protocol, the maximum coverage range is limited to what the underlying IEEE 802.15.4 technology can provide [83]. The maximum coverage range of the IEEE 802.15.4 technology is up to 100 m [79]. This range is the range between two nodes; as the topology of a Thread network is mesh, the actual coverage and size of the network could span greater than 100 m. The maximum throughput is also restricted to the specification of the IEEE 802.15.4 technology, with a quoted maximum of 250 kbps [84]. As with Zigbee, the maximum frame length of the underlying IEEE 802.15.4 technology is 128 bytes [81].

Herrera and Núñez [118] developed a model Thread network to simulate a HAN. They found that the network could operate with a low packet loss rate; however, the border router may not be able to cope efficiently in times of high load. This may be mitigated by having multiple border routers; however, further study is required [118].

Thread uses multiple mechanisms to ensure security. There is frame security implemented based on IEEE 802.15.4 [85]. The Mesh Commissioning Protocol applies security when new devices join the network [85]. Thread can also support AES-128 for encryption; however, due to scalability issues, Elliptic Curve Cryptography (ECC) is used [85]. Thread is a newer protocol with a promising future endorsed by major industry players. Thread is a candidate to support the communication needs of the HAN.

3.7 Wi-Fi

Wi-Fi is the name for a wide range of wireless network protocols that conform to the numerous and ever-growing family of IEEE 802.11 standards. The Wi-Fi Alliance holds the rights to the Wi-Fi name and is responsible for certifying products and companies that comply with the various Wi-Fi standards [119]. Wi-Fi has become ubiquitous and is a popular choice for use in the LAN and HAN [111]. Wi-Fi is a mature technology, and as such, there have been numerous iterations over the years. In the latest released specification, IEEE 802.11ax [120], Speeds can exceed 1 Gbps [86], making it a suitable choice for high bandwidth applications. Modern Wi-Fi standards use a frame aggregation process that combines several smaller frames into a larger one for transmission [121]. This is known as an aggregate MAC protocol data unit (A-MDPU). The maximum size of an A-MDPU in an IEEE 802.11ax-based Wi-Fi network is 4,194,304 bytes [87]. In modern Wi-Fi networks, confidentiality authentication and access control are typically provided by either Wi-Fi Protected Access II/III (WPA2/WPA3) [88]. Work is underway by the IEEE 802.11be task group to bring substantial performance improvements to the next generation of Wi-Fi by increasing throughput and reducing latency [122].

A promising new technology in the Wi-Fi space is IEEE 802.11ah (also known as Wi-Fi HaLow) [123], a recent addition to the IEEE 802.11 (Wi-Fi) standard explicitly aimed at the communication needs of IoT devices, particularly wireless sensor networks [89]. It operates in the Sub-GHz bands, typically somewhere in the 900 MHz range where the spectrum is unlicensed [89]. It can support long-range transmission up to 1 km in an outdoor environment [79, 89] with a theoretical maximum speed of 86.7 Mbps [90]. The network topology of IEEE 802.11ah is single-hop [79, 89] and has an A-MDPU of 7991 bytes [91]. Shorter headers have been implemented to reduce unnecessary data transmission [79, 89]. Wi-Fi HaLow uses Wi-Fi Protected Access 3 (WPA3) to secure communications between end devices and access points. It includes additional security capabilities to secure communications [123, 124]. One of the enhancements includes the implementation of Simultaneous Authentication of Equals (SAE) which aims to stop dictionary attacks [125].

As an emerging technology with no current practical implementations, 802.11ah shows exciting potential as a technology to enable communication in the NAN. This is due to its promised high throughput, compatibility with existing well-understood security standards, and extended network coverage over traditional Wi-Fi.

3.8 SigFox

SigFox is an unlicensed band, sub-GHz LPWAN technology. Like LoRa, it is also subject to duty cycle restrictions that limit the frequency of communications allowed to be transmitted depending on geographical location [93]. SigFox is suitable for low-powered applications due to its efficient Differential Binary Phase-Shift Keying

(D-BPSK) modulation technology [93]. SigFox transmissions are limited by a data rate of 100 bps in Europe and 600 bps in the United States [2, 93]. The maximum payload that can be included in a SigFox packet at any time is only 12 bytes [93]. SigFox can support long-range communications up to 40 km in a rural setting and up to 10 km in an urban environment [92].

SigFox has a built-in scheme to ensure data integrity and authenticity. A unique key is burnt-in to each SigFox device when it is manufactured that is used in this process [126]. The authentication scheme uses the AES-128 cipher and a network access key [94].

While SigFox utilises the free, unlicensed band, subscription charges apply to access its "0G" network [127]. This introduces costs that may not be incurred with other unlicensed band technologies. Confidentiality protection through encryption is optional, although the details of how it works are somewhat vague [94].

Due to its long-range capability, SigFox can cover great distances with minimal base stations, making it a candidate for consideration for the NAN and WAN; however, due to the limited payload and the duty cycle restrictions, SigFox is unsuitable for applications that need to transmit frequently or moderate to large packets [92].

3.9 Narrowband-IoT (NB-IoT)

NB-IoT is a licensed band cellular technology maintained by the 3GPP. Unlike some of the other mentioned unlicensed band IoT transmission technologies, the use of the spectrum has a financial cost involved. NB-IoT is highly scalable and can support up to 100,000 devices in any given cell [92]. This contrasts with IoT transmission technologies, such as LoRa and SigFox, which only support half that amount [92]. As NB-IoT uses LTE (Long Term Evolution) cellular networks, it is only suitable where these existing networks already exist [92]. In situations with poor LTE coverage, NB-IoT is not a technology that should be considered for deployment. NB-IoT is suitable for applications with low latency requirements; however, it is not as energy efficient as other LPWAN technologies [92].

NB-IoT supports data rates up to 200 kbps [95]. The supported payload size is far greater than other LPWAN technologies offering up to 1600 bytes per transmission [92]. Whilst the coverage range of NB-IoT is up to 10 km in a rural setting, this drops substantially in an urban environment, delivering up to 1 km [95].

Because NB-IoT is derived from LTE, it also shares its security architecture [96]. NB-IoT supports mutual authentication between the end device and the network and provides confidentiality using algorithms such as ATR-128 and SNOW 3G [96]. Each session is protected with an HMAC-SHA256 session key [96].

3.10 LTE-M

LTE-M (Long Term Evolution for Machines) is a licensed band technology maintained by the 3GPP (3rd Generation Partnership Project). LTE-M is related to the 4G LTE technology commonly used in cellular phone networks [128, 129]. It is also closely related to the NB-IoT technology; however, LTE-M can provide better throughput (1 Mbps uplink and downlink) while delivering lower latency [128]. Whilst the data rates are more attractive than some of the unlicensed band technologies that have been discussed, LTE-M being a licensed band transmission technology, requires an ongoing subscription from a carrier.

While the maximum data rate LTE-M provides is significantly higher than NB-IoT at up to 1 Mbps, the maximum coverage that can be provided is lower at around 5 km [97]. The maximum payload for a transmission can vary based on numerous factors, such as the network operator, so it is recommended to keep the entire packet under 1280 bytes [98]. Like NB-IoT, LTE-M being an LTE-based technology also shares the security architecture [130].

3.11 5G

The rollout of 5G cellular technology is still in the early days. The world's first commercially available cellular network for smartphones was launched in 2019 [131, 132]. 5G promises faster data rates and lower latency over existing 4G LTE technologies. There are many advancements to come for 5G in the IoT space, with many research challenges to be addressed. According to Chettri and Bera [133], the current research focuses on increasing data rates over existing technologies, providing lower latency and lower-cost networks that can support more devices while conserving battery power through reduced energy consumption.

4 Conclusion

IoT security has never been more critical as we continue to integrate more physical systems with data networks and the Internet. This is particularly true when discussing critical infrastructure such as the electricity grid. This so-called 'smartification' of the grid with IoT technology enables many possibilities, such as improved visibility of network outages, cost reduction benefits, efficient billing practices, improved safety, and improved environmental outcomes realised through real-time data. It is, however, not without its challenges.

This chapter has discussed the various data networks that make up the IoT-based smart grid, the HAN, the NAN, and the WAN. We have covered the purpose of each network and a summary of some of the security-related research in each of these networks.

We then covered security research in the context of the HAN, NAN, and WAN. Attacks that risk the stability of the IoT-smart grid were outlined and summarised. Security challenges that exist in the smart grid were then discussed. The IoT-smart grid faces many challenges in encryption, authentication, authorisation, and auditing. Safety and privacy concerns relating to the smart grid were also examined. Some key future research areas were identified from the literature, such as the need for new protocols and frameworks to support the smart grid's current and future security requirements.

A selection of modern IoT transmission technologies was surveyed. Their key performance areas were discussed, including throughput, coverage, cost, duty cycle, maximum payload, and built-in security measures.

It is essential to choose the right mix of technologies with the proper security protections and adequate performance measures to ensure the networks have sufficient performance for the HAN/NAN/WAN. It is also critical that potential threats be minimised in these critical networks.

Acknowledgements This work has been supported by the Cyber Security Research Centre Limited whose activities are partially funded by the Australian Government's Cooperative Research Centres Programme.

References

1. Al-Turjman F, Abujubbeh M (2019) IoT-enabled smart grid via SM: an overview. Future Gener Comput Syst 96:579–590
2. Lavric A, Petrariu AI, Popa V (2019) SigFox communication protocol: the new era of IoT? In: 2019 international conference on sensing and instrumentation in IoT era (ISSI). IEEE
3. Haddad Z et al (2015) Secure and privacy-preserving AMI-utility communications via LTE-A networks. In: 2015 IEEE 11th international conference on wireless and mobile computing, networking and communications (WiMob). IEEE.
4. Talha SK, Barry BI (2013) Evaluating the impact of AES encryption algorithm on Voice over Internet Protocol (VoIP) systems. In: 2013 international conference on computing, electrical and electronic engineering (ICCEEE). IEEE
5. U.S. Department of Energy (2022) The smart grid. Cited 31 Aug 2022. Available from: https://www.smartgrid.gov/the_smart_grid/smart_grid.html
6. Nidhi N, Prasad D, Nath V (2019) Different aspects of smart grid: an overview. In: Nanoelectronics, circuits and communication systems, pp 451–456
7. Butt OM, Zulqarnain M, Butt TM (2021) Recent advancement in smart grid technology: future prospects in the electrical power network. Ain Shams Eng J 12(1):687–695
8. Kim SC, Ray P, Reddy SS (2019) Features of smart grid technologies: an overview. ECTI Trans Electr Eng Electron Commun 17(2):169–180
9. Yoldaş Y et al (2017) Enhancing smart grid with microgrids: challenges and opportunities. Renew Sustain Energy Rev 72:205–214
10. Gunduz MZ, Das R (2020) Cyber-security on smart grid: threats and potential solutions. Comput Netw 169:107094

11. Straub J (2021) Consideration of the use of smart grid cyberattacks as an influence attack and appropriate deterrence. In: 2021 international conference on computational science and computational intelligence (CSCI). IEEE
12. Ahmed S et al (2019) A survey on communication technologies in smart grid. In: 2019 IEEE PES GTD grand international conference and exposition Asia (GTD Asia). IEEE
13. Nejat P et al (2015) A global review of energy consumption, CO_2 emissions and policy in the residential sector (with an overview of the top ten CO_2 emitting countries). Renew Sustain Energy Rev 43:843–862
14. Mendes TD et al (2015) Smart home communication technologies and applications: wireless protocol assessment for home area network resources. Energies 8(7):7279–7311
15. Hu Q, Li F (2013) Hardware design of smart home energy management system with dynamic price response. IEEE Trans Smart Grid 4(4):1878–1887
16. Jo H-C, Kim S, Joo S-K (2013) Smart heating and air conditioning scheduling method incorporating customer convenience for home energy management system. IEEE Trans Consum Electron 59(2):316–322
17. Zhou B et al (2016) Smart home energy management systems: concept, configurations, and scheduling strategies. Renew Sustain Energy Rev 61:30–40
18. Joseph S, Menon DM (2015) A novel architecture for efficient communication in smart grid home area network. In: 2015 IEEE international conference on computational intelligence and computing research (ICCIC). IEEE
19. Agnew S, Dargusch P (2017) Consumer preferences for household-level battery energy storage. Renew Sustain Energy Rev 75:609–617
20. Kumar MS, Srinivasan S, Subathra B (2018) A deterministic demand response program for schedulable loads in power distribution system. Int J Pure Appl Math 118(18):2071–2077
21. Alimi OA, Ouahada K (2018) Security assessment of the smart grid: a review focusing on the NAN architecture. In: 2018 IEEE 7th international conference on adaptive science & technology (ICAST). IEEE
22. Nafi NS et al (2018) Software defined neighborhood area network for smart grid applications. Future Gener Comput Syst 79:500–513
23. Ramirez DF et al (2015) Performance evaluation of future AMI applications in smart grid neighborhood area networks. In: IEEE Colombian conference on communication and computing (IEEE COLCOM 2015). IEEE
24. Kulkarni V, Komanapalli VLN, Sahoo SK (2021) A review on requirements for data communication and information technology areas for smart grid. In: Advances in automation, signal processing, instrumentation, and control, pp 3259–3271
25. Kuzlu M, Pipattanasomporn M, Rahman S (2014) Communication network requirements for major smart grid applications in HAN, NAN and WAN. Comput Netw 67:74–88
26. Pandey JC, Kalra M (2022) A review of security concerns in smart grid. In: Innovative data communication technologies and application, pp 125–140
27. Kumar NM, Mallick PK (2018) The Internet of Things: insights into the building blocks, component interactions, and architecture layers. Procedia Comput Sci 132:109–117
28. Li L (2012) Study on security architecture in the Internet of Things. In: Proceedings of 2012 international conference on measurement, information and control. IEEE
29. Lee S, Kim J, Shon T (2016) User privacy-enhanced security architecture for home area network of smartgrid. Multimed Tools Appl 75(20):12749–12764
30. Haes Alhelou H et al (2019) A survey on power system blackout and cascading events: research motivations and challenges. Energies 12(4):682
31. Akatyev N, James JI (2019) Evidence identification in IoT networks based on threat assessment. Future Gener Comput Syst 93:814–821
32. Menon DM, Radhika N (2016) Anomaly detection in smart grid traffic data for home area network. In: 2016 international conference on circuit, power and computing technologies (ICCPCT). IEEE
33. Ali W et al (2022) A novel privacy preserving scheme for smart grid-based home area networks. Sensors 22(6):2269

34. de Melo PH, Miani RS, Rosa PF (2022) FamilyGuard: a security architecture for anomaly detection in home networks. Sensors 22(8):2895
35. Menon DM, Radhika N. A trust-based framework and deep learning-based attack detection for smart grid home area network
36. Holman BA, Hauser J, Amariucai GT (2021) Toward home area network hygiene: device classification and intrusion detection for encrypted communications. In: Advances in security, networks, and Internet of Things. Springer, pp 195–209
37. Mendel J (2017) Smart grid cyber security challenges: overview and classification. e-mentor 1(68):55–66
38. Kaveh M, Mosavi MR (2020) A lightweight mutual authentication for smart grid neighborhood area network communications based on physically unclonable function. IEEE Syst J 14(3):4535–4544
39. Yilmaz Y, Uludag S (2017) Mitigating IoT-based cyberattacks on the smart grid. In: 2017 16th IEEE international conference on machine learning and applications (ICMLA). IEEE
40. Kalidass J, Purusothaman T, Suresh P (2021) Enhancement of end-to-end security in advanced metering infrastructure. J Ambient Intell Humaniz Comput 1–10
41. Alohali B et al (2016) Group authentication scheme for neighbourhood area networks (NANs) in smart grids. J Sens Actuator Netw 5(2):9
42. Rawat DB, Bajracharya C (2015) Cyber security for smart grid systems: status, challenges and perspectives. In: SoutheastCon 2015, pp 1–6
43. Kharchouf I et al (2022) On the implementation and security analysis of routable-GOOSE messages based on IEC 61850 standard. In: 2022 IEEE international conference on environment and electrical engineering and 2022 IEEE industrial and commercial power systems Europe (EEEIC/I&CPS Europe). IEEE
44. Islam SN, Baig Z, Zeadally S (2019) Physical layer security for the smart grid: vulnerabilities, threats, and countermeasures. IEEE Trans Ind Inform 15(12):6522–6530
45. Komninos N, Philippou E, Pitsillides A (2014) Survey in smart grid and smart home security: issues, challenges and countermeasures. IEEE Commun Surv Tutor 16(4):1933–1954
46. Liang X, Kim Y (2021) A survey on security attacks and solutions in the IoT network. In: 2021 IEEE 11th annual computing and communication workshop and conference (CCWC). IEEE
47. Mathas C-M et al (2020) Threat landscape for smart grid systems. In: Proceedings of the 15th international conference on availability, reliability and security
48. Chhaya L et al (2020) Cybersecurity for smart grid: threats, solutions and standardization. In: Advances in greener energy technologies. Springer, pp 17–29
49. Tufail S et al (2021) A survey on cybersecurity challenges, detection, and mitigation techniques for the smart grid. Energies 14(18):5894
50. Yılmaz Y, Uludag S (2021) Timely detection and mitigation of IoT-based cyberattacks in the smart grid. J Franklin Inst 358(1):172–192
51. Imteaj A et al (2021) A survey on federated learning for resource-constrained IoT devices. IEEE Internet Things J 9(1):1–24
52. Alam S et al (2020) Internet of Things (IoT) enabling technologies, requirements, and security challenges. In: Advances in data and information sciences. Springer, pp 119–126
53. Singh S et al (2017) Advanced lightweight encryption algorithms for IoT devices: survey, challenges and solutions. J Ambient Intell Humaniz Comput 1–18
54. Fernández-Caramés TM (2019) From pre-quantum to post-quantum IoT security: a survey on quantum-resistant cryptosystems for the Internet of Things. IEEE Internet Things J 7(7):6457–6480
55. Bekara C (2014) Security issues and challenges for the IoT-based smart grid. Procedia Comput Sci 34:532–537
56. Arakadakis K et al (2021) Firmware over-the-air programming techniques for IoT networks—a survey. ACM Comput Surv (CSUR) 54(9):1–36
57. Malhotra P et al (2021) Internet of Things: evolution, concerns and security challenges. Sensors 21(5):1809

58. Raghuvanshi A et al (2021) An investigation of various applications and related security challenges of Internet of Things. Mater Today Proc
59. Gopstein A et al (2021) NIST framework and roadmap for smart grid interoperability standards, release 4.0. Department of Commerce, National Institute of Standards and Technology
60. Karale A (2021) The challenges of IoT addressing security, ethics, privacy, and laws. Internet Things 15:100420
61. Kimani K, Oduol V, Langat K (2019) Cyber security challenges for IoT-based smart grid networks. Int J Crit Infrastruct Prot 25:36–49
62. Saleem Y et al (2019) Internet of Things-aided smart grid: technologies, architectures, applications, prototypes, and future research directions. IEEE Access 7:62962–63003
63. Sanchez-Gomez J, Sanchez-Iborra R, Skarmeta A (2017) Transmission technologies comparison for IoT communications in smart-cities. In: GLOBECOM 2017–2017 IEEE global communications conference. IEEE
64. Campo GD et al (2018) Power distribution monitoring using LoRa: coverage analysis in suburban areas. In: Proceedings of the 2018 international conference on embedded wireless systems and networks
65. Seye MR, Gueye B, Diallo M (2017) An evaluation of LoRa coverage in Dakar Peninsula. In: 2017 8th IEEE annual information technology, electronics and mobile communication conference (IEMCON). IEEE
66. Adelantado F et al (2017) Understanding the limits of LoRaWAN. IEEE Commun Mag 55(9):34–40
67. Noreen U, Bounceur A, Clavier L (2017) A study of LoRa low power and wide area network technology. In: 2017 international conference on advanced technologies for signal and image processing (ATSIP). IEEE
68. Mekki K et al (2018) Overview of cellular LPWAN technologies for IoT deployment: SigFox, LoRaWAN, and NB-IoT. In: 2018 IEEE international conference on pervasive computing and communications workshops (PerCom workshops). IEEE
69. LoRa Alliance (2017) LoRaWAN™ security full end-to-end encryption for IoT application providers. Cited 1 May 2022. Available from: https://lora-alliance.org/sites/default/files/2019-05/lorawan_security_whitepaper.pdf
70. Janssen T et al (2020) LoRa 2.4 GHz communication link and range. Sensors 20(16):4366
71. Andersen FR et al (2020) Ranging capabilities of LoRa 2.4 GHz. In: 2020 IEEE 6th world forum on Internet of Things (WF-IoT). IEEE
72. Semtech (2020) SX1280/SX1281 data sheet DS.SX1280-1.W.APP Rev 3.2
73. Mulla A et al (2015) The wireless technologies for smart grid communication: a review. In: 2015 fifth international conference on communication systems and network technologies. IEEE
74. Bluetooth SIG (2022) Learn about Bluetooth—Bluetooth technology overview. Cited 11 Sept 2022. Available from: https://www.bluetooth.com/learn-about-bluetooth/tech-overview/
75. T'Jonck K et al (2021) Optimizing the Bluetooth low energy service discovery process. Sensors 21(11):3812
76. Bluetooth SIG (2020) Topology options | Bluetooth® technology website. Cited 11 Apr 2020. Available from: https://www.bluetooth.com/learn-about-bluetooth/bluetooth-technology/topology-options/
77. Yin J et al (2019) A survey on Bluetooth 5.0 and mesh: new milestones of IoT. ACM Trans Sens Netw (TOSN) 15(3):1–29
78. Anupriya K, Yomas J, Dwarakanath T (2016) Integrating ZigBee and sub GHz devices for long range networks. In: 2016 online international conference on green engineering and technologies (IC-GET). IEEE
79. Ahmed N, Rahman H, Hussain MI (2016) A comparison of 802.11ah and 802.15.4 for IoT. ICT Express 2(3):100–102
80. Kumar T, Mane P (2016) ZigBee topology: a survey. In: 2016 international conference on control, instrumentation, communication and computational technologies (ICCICCT). IEEE

81. Brachmann M et al (2019) IEEE 802.15.4 TSCH in sub-GHz: design considerations and multi-band support. In: 2019 IEEE 44th conference on local computer networks (LCN). IEEE
82. Datta P, Sharma B (2017) A survey on IoT architectures, protocols, security and smart city based applications. In: 2017 8th international conference on computing, communication and networking technologies (ICCCNT). IEEE
83. Herrero R (2022) Thread architecture. In: Fundamentals of IoT communication technologies. Springer, pp 213–225
84. Rzepecki W, Iwanecki Ł, Ryba P (2018) IEEE 802.15.4 thread mesh network–data transmission in harsh environment. In: 2018 6th international conference on future Internet of Things and cloud workshops (FiCloudW). IEEE
85. Unwala I, Taqvi Z, Lu J (2018) Thread: an IoT protocol. In: 2018 IEEE green technologies conference (GreenTech). IEEE
86. Edirisinghe S, Wijethunge A, Ranaweera C. Wi-Fi 6-based HAN for demand response in smart grid. Available at SSRN 4184450
87. Sharon O, Alpert Y (2017) Single user MAC level throughput comparison: IEEE 802.11ax vs. IEEE 802.11ac. Wireless Sens Netw 9(5):166–177
88. Yang G (2022) An overview of current solutions for privacy in the Internet of Things. Front Artif Intell 5
89. Adame T et al (2014) IEEE 802.11ah: the WiFi approach for M2M communications. IEEE Wireless Commun 21(6):144–152
90. Wi-Fi Alliance (2021) Wi-Fi certified HaLow technology overview. Cited 9 Sept 2022. Available from: https://www.wi-fi.org/downloads-registered-guest/Wi-Fi_CERTIFIED_HaLow_Technology_Overview_20211102.pdf/36879
91. Vega LFDC, Robles I, Morabito R (2015) IPv6 over 802.11ah. Internet Engineering Task Force, p 17
92. Mekki K et al (2019) A comparative study of LPWAN technologies for large-scale IoT deployment. ICT Express 5(1):1–7
93. Lavric A, Petrariu AI, Popa V (2019) Long range SigFox communication protocol scalability analysis under large-scale, high-density conditions. IEEE Access 7:35816–35825
94. Ferreira L (2021) Sigforgery: breaking and fixing data authenticity in SigFox. In: International conference on financial cryptography and data security. Springer
95. Lalle Y et al (2019) A comparative study of LoRaWAN, SigFox, and NB-IoT for smart water grid. In: 2019 global information infrastructure and networking symposium (GIIS). IEEE
96. Nair KK, Abu-Mahfouz AM, Lefophane S (2019) Analysis of the narrow band Internet of Things (NB-IoT) technology. In: 2019 conference on information communications technology and society (ICTAS). IEEE
97. Aldahdouh KA, Darabkh KA, Al-Sit W (2019) A survey of 5G emerging wireless technologies featuring LoRaWAN, SigFox, NB-IoT and LTE-M. In: 2019 international conference on wireless communications signal processing and networking (WiSPNET). IEEE
98. Amazon Web Services (2022) Implementing low-power wide-area network (LPWAN) solutions with AWS IoT. AWS whitepaper. Cited 11 Sept 2022. Available from: https://docs.aws.amazon.com/whitepapers/latest/implementing-lpwan-solutions-with-aws/lte-m.html
99. Augustin A et al (2016) A study of LoRa: long range & low power networks for the Internet of Things. Sensors 16(9):1466
100. de Almeida IBF et al (2021) Alternative chirp spread spectrum techniques for LPWANs. IEEE Trans Green Commun Netw 5(4):1846–1855
101. Angrisani L et al (2017) LoRa protocol performance assessment in critical noise conditions. In: 2017 IEEE 3rd international forum on research and technologies for society and industry (RTSI). IEEE
102. Magrin D, Centenaro M, Vangelista L (2017) Performance evaluation of LoRa networks in a smart city scenario. In: 2017 IEEE international conference on communications (ICC). IEEE
103. Zorbas D et al (2018) Improving LoRa network capacity using multiple spreading factor configurations. In: 2018 25th international conference on telecommunications (ICT). IEEE

104. The Things Network (2022) Spreading factors. Cited 9 June 2022. Available from: https://www.thethingsnetwork.org/docs/lorawan/spreading-factors/
105. LoRa Alliance (2022) What is LoRaWAN specification. Cited 1 May 2022. Available from: https://lora-alliance.org/about-lorawan/
106. del Campo G et al (2018) Power distribution monitoring using LoRa: coverage analysis in suburban areas
107. Polak L, Milos J (2020) Performance analysis of LoRa in the 2.4 GHz ISM band: coexistence issues with Wi-Fi. Telecommun Syst 1–11
108. Semtech (2020) Semtech SX1280 datasheet Rev 3.2. Cited 31 May 2021. Available from: https://www.semtech.com/products/wireless-rf/lora-24ghz/sx1280#download-resources
109. Collotta M et al (2018) Bluetooth 5: a concrete step forward toward the IoT. IEEE Commun Mag 56(7):125–131
110. Bluetooth SIG (2021) Bluetooth core specification v5.3
111. Mahmood A, Javaid N, Razzaq S (2015) A review of wireless communications for smart grid. Renew Sustain Energy Rev 41:248–260
112. Krejčí R, Hujňák O, Švepeš M (2017) Security survey of the IoT wireless protocols. In: 2017 25th telecommunication forum (TELFOR). IEEE
113. IEEE Standard for Low-Rate Wireless Networks (2016) IEEE Std 802.15.4-2015 (revision of IEEE Std 802.15.4-2011), pp 1–709
114. Ali AI et al (2019) ZigBee and LoRa based wireless sensors for smart environment and IoT applications. In: 2019 1st global power, energy and communication conference (GPECOM). IEEE
115. Thread Group (2022) Thread 1.3.0 features white paper. Cited 18 Aug 2022. Available from: https://www.threadgroup.org/Portals/0/documents/support/Thread1.3.0WhitePaper_07192022_3990_1.pdf
116. Google (2020) OpenThread. Cited 11 Apr 2020. Available from: https://github.com/openthread/openthread
117. Cilfone A et al (2019) Wireless mesh networking: an IoT-oriented perspective survey on relevant technologies. Future Internet 11(4):99
118. Herrera T, Núñez F (2019) Scalability and integration of a thread implementation in a home area network. In: 2019 IEEE international conference on consumer electronics (ICCE). IEEE
119. Wi-Fi Alliance (2022) Who we are—our brands. Cited 12 Aug 2022. Available from: https://www.wi-fi.org/who-we-are/our-brands
120. IEEE Standard for Information Technology—Telecommunications and Information Exchange Between Systems Local and Metropolitan Area Networks—Specific Requirements Part 11: Wireless LAN Medium Access Control (MAC) and Physical Layer (PHY) Specifications Amendment 1 (2021) Enhancements for high-efficiency WLAN. IEEE Std 802.11ax-2021 (amendment to IEEE Std 802.11-2020), pp 1–767
121. Baños-Gonzalez V et al (2016) IEEE 802.11ah: a technology to face the IoT challenge. Sensors 16(11):1960
122. Adame T, Carrascosa-Zamacois M, Bellalta B (2021) Time-sensitive networking in IEEE 802.11be: on the way to low-latency WiFi 7. Sensors 21(15):4954
123. Wi-Fi Alliance (2018) Next generation Wi-Fi®: the future of connectivity. Cited 8 Apr 2020. Available from: https://www.wi-fi.org/download.php?file=/sites/default/files/private/Next_generation_Wi-Fi_White_Paper_20181218.pdf
124. Seferagić A et al (2020) Survey on wireless technology trade-offs for the industrial Internet of Things. Sensors 20(2):488
125. Vanhoef M, Ronen E (2019) Dragonblood: a security analysis of WPA3's SAE handshake. IACR Cryptol ePrint Arch 2019:383
126. Chacko S, Job MD (2018) Security mechanisms and vulnerabilities in LPWAN. IOP Conf Ser Mater Sci Eng. IOP Publishing
127. SigFox (2022) Buy SigFox connectivity for your IoT devices. Cited 17 Aug 2022. Available from: https://buy.sigfox.com

128. Borkar SR (2020) Long-term evolution for machines (LTE-M). In: LPWAN technologies for IoT and M2M applications. Elsevier, pp 145–166
129. Dian FJ, Vahidnia R (2020) LTE IoT technology enhancements and case studies. IEEE Consum Electron Mag 9(6):49–56
130. Ugwuanyi S, Irvine J (2020) Security analysis of IoT networks and platforms. In: 2020 international symposium on networks, computers and communications (ISNCC). IEEE
131. McCurry J (2019) US dismisses South Korea's launch of world-first 5G network as 'stunt'. The Guardian. Cited 28 Aug 2022. Available from: https://www.theguardian.com/technology/2019/apr/04/us-dismisses-south-koreas-launch-of-world-first-5g-network-as-stunt
132. Su J (2019) Verizon launches world's first commercial 5G smartphone service. Forbes. Cited 28 Aug 2022. Available from: https://www.forbes.com/sites/jeanbaptiste/2019/04/04/verizon-launches-worlds-first-commercial-5g-smartphone-service/?sh=32c3fa001961
133. Chettri L, Bera R (2019) A comprehensive survey on Internet of Things (IoT) toward 5G wireless systems. IEEE Internet Things J 7(1):16–32

MIX
Papier aus verantwortungsvollen Quellen
Paper from responsible sources
FSC® C105338

If you have any concerns about our products,
you can contact us on
ProductSafety@springernature.com

In case Publisher is established outside the EU,
the EU authorized representative is:
Springer Nature Customer Service Center GmbH
Europaplatz 3, 69115 Heidelberg, Germany

Printed by Libri Plureos GmbH
in Hamburg, Germany